'On the face of it, Roger Leakey's contention – that, through the careful integration of trees on farms, there is more than enough capacity to produce food to meet the needs of a growing world population – is a bold one. But in this very readable volume, which dovetails Roger's accumulated wisdom from a distinguished research career with his barely disguised passion to improve the lot of poor smallholders worldwide, he demonstrates convincingly that it actually can be done. Read it, believe it and pass the news on.'
Mike Turnbull, Chairman, International Tree Foundation, Crawley Down, UK

'A fine, wise and enormously important book about trees and people, showing how we can live better by redesigning agricultural systems. More sustainable systems can work, and this book draws on evidence to show how production systems can be good for both people and the planet.'
Jules Pretty, Deputy Vice-Chancellor of Science & Engineering and Sustainability & Resources, University of Essex, UK

'*Living with the Trees of Life* presents practical, common sense solutions that will uplift and empower farmers, educators, assistance providers, and policymakers.'
Craig Elevitch, Director, Agroforestry Net, Hawaii, USA

'Roger is something of a visionary, and this inspirational book presents powerful evidence of what can be achieved through a lifetime's dedication, hard work and by building a multidisciplinary team.'
Prof. Adrian Newton, Professor of Conservation Science, University of Bournemouth, UK

'Part personal journey, part scientific biography, this book charts the evolution of agroforestry from an under-researched traditional farming practice to an interdisciplinary and transformative approach to agriculture…Read it and be inspired!'
Dr Kate Schreckenberg, Coordinator, Centre for Underutilised Crops, University of Southampton, Southampton, UK

'If you read only one book this year about the challenges facing global society, this is the one for you!'
Dr Charles Clement, National Research Institute for Amazonia, Manaus, Amazonas, Brazil

'A must-read for those who take sustainable food security provision in the tropics seriously.'
Prof. Patrick van Damme, Plant Production Department, Tropical and Sub-tropical Agriculture and Ethnobotany Laboratory, University of Ghent, Belgium

'We cannot afford to ignore the principal message that unfolds as a legacy of Roger's rich experience in agroforestry: that we can empower the peoples of developing tropical economies with productive and socially and environmentally sustainable strategies to ensure a brighter future for all.'
Prof. Paul Gadek, Centre for Tropical Agri-Tech Research, James Cook University, Cairns, Australia

'This account is inspiring and thought-provoking both for the student and the seasoned practitioner of sustainable land use and agricultural development. It is also an excellent introduction for the interested layperson.'
Dr Goetz Schroth, Federal University of Western Pará, Santarém, Brazil

'There is a growing appreciation for the value of agroforestry, and this book will contribute to the wealth of knowledge needed by a variety of practitioners – from farmers, teachers, researchers, and policymakers.'
Prof. Judi Wakhungu, Executive Director, African Centre for Technology Studies, Nairobi, Kenya

'This book could not have come at a more opportune time. Leakey's knowledge, deep wisdom, scientific expertise and long years in the service of smallholder farmers, and of the "trees of life" that sustain them, make him the ideal storyteller to show how science can be melded with traditional knowledge to develop workable agroforestry solutions to the many crises that confront life on earth. This is a book that can truly help the "bottom billion".'
Joan Baxter, Senior Research Fellow, Oakland Institute, California, USA

'Roger Leakey's book considers the huge challenges for the poorer nations of the world and the responsibility of developed countries to engage in the debates about sustainable food security, competing uses for land, and the development of new resources to meet the demands of local communities for food and fuel.'
Fiona O'Donnell, Member of Parliament for East Lothian, House of Commons, London, UK

'I cannot think of any better person to write this book, which brings new understanding to pervasive problems across the human–environment–development interface. There is no doubt that this book will become a seminal text for agricultural, development and environmental planners, policy-makers and practitioners throughout the world.'
Prof. Charlie Shackleton, Head of Environmental Science, Rhodes University, South Africa

'Although I am a forest tree geneticist engaged in breeding for industrial tree plantations, my long association with agroforestry and participatory tree domestication has convinced me of their central contribution to paths out of poverty for the rural tropical poor, and for a sustainable world.'
Dr Chris Harwood, CSIRO, Tasmania, Australia

Living with the Trees of Life

Towards the Transformation of Tropical Agriculture

To Alison, Andrew and Chris

Front cover image: Two farmers, FOUDA Bernard and MAMOLE Angeline from Ngali II (Centre Region of Cameroon), examining fruits of safou (*Dacryodes edulis*) or African plum, a highly nutritious and widely traded local food.

Living with the Trees of Life
Towards the Transformation of Tropical Agriculture

Roger R.B. Leakey

www.cabi.org

CABI is a trading name of CAB International

CABI	CABI
Nosworthy Way	875 Massachusetts Avenue
Wallingford	7th Floor
Oxfordshire OX10 8DE	Cambridge, MA 02139
UK	USA
Tel: +44 (0)1491 832111	Tel: +1 617 395 4056
Fax: +44 (0)1491 833508	Fax: +1 617 354 6875
E-mail: cabi@cabi.org	E-mail: cabi-nao@cabi.org
Website: www.cabi.org	

Copyright © 2012 Roger Leakey. All rights reserved. No part of this publication may be reproduced in any form or by any means, electronically, mechanically, by photocopying, recording or otherwise, without the prior permission of the copyright owners.

A catalogue record for this book is available from the British Library, London, UK.

Library of Congress Cataloging-in-Publication Data

Leakey, Roger R. B.
  Living with the trees of life: towards the transformation of tropical agriculture / [Roger Leakey].
    p. cm.
  Includes index.
  ISBN 978-1-78064-099-0 (hardback)
1. Tree crops--Tropics. 2. Agroforestry--Tropics. I. Title.

SB171.T73L43 2012
634--dc23
                            2012006161

ISBN-13: 978 1 78064 099 0 (HBK)
         978 1 78064 098 3 (PBK)

Commissioning editor: Claire Parfitt
Editorial assistant: Chris Shire
Production editor: Tracy Head

Typeset by SPi, Pondicherry, India.
Printed and bound in the UK by MPG Books Ltd.

Contents

About the Author — ix

Foreword — xi

Preface — xiii

Acknowledgements — xix

Frequently Used Acronyms — xxi

1. Revelations in Kumba — 1
2. The Big Global Issues — 13
3. Journeys of Discovery in Agroforestry — 24
4. Diversity and Function in Farming Systems — 51
5. Finding the Trees of Life — 65
6. Selecting the Best Trees — 83
7. Vegetative Propagation — 95
8. Case Studies from the Pacific — 112
9. Marketing Tree Products — 125
10. Redirecting Agriculture – Going Multifunctional — 141
11. Multifunctional Agriculture – Proof of Concept — 156
12. The Convenient Truths — 170

Postscript — 180

Appendix: Author's Experience Prior to the Events of this Book — 181

Index — 185

About the Author

Professor Roger Leakey DSc, PhD, BSc, NDA is a former Director of Research at the International Centre for Research in Agroforestry (ICRAF, 1993–1997) and Professor of Agroecology and Sustainable Development of James Cook University, Cairns, Australia (2001–2006). He is Vice President of the International Society of Tropical Foresters and is Vice Chairman of the International Tree Foundation. He holds a number of fellowships in learned societies, universities and international research centres. He was a coordinating lead author in the International Assessment of

Agricultural Science and Technology for Development (IAASTD), which was approved by 58 governments in an intergovernmental plenary meeting in Johannesburg, South Africa in April 2008. This assessment examined the impact of agricultural knowledge, science and technology on environmentally, socially and economically sustainable development worldwide over the last 50 years, and suggested that to meet these challenges agriculture has to advance from a unifunctional focus on food production and to additionally embrace more environmental, social and economic goals – i.e. to become multifunctional.

To advance agriculture in this direction, the author initiated what has become a global programme to start the domestication of wild fruit and nut trees that were the staple diet of people before the Green Revolution raised the profile of a few starch crops. This involved the development of some robust horticultural techniques that can be implemented in remote corners of the developing world, as well as some basic studies of the biology of potential food crops that are unknown to most of us.

This book presents the story of these changes in agricultural philosophy within the context of the author's personal experience of travelling and working in many countries of North, Central and South America and the Caribbean, Africa, the Middle East, South and South-east Asia and Oceania.

Foreword

How the world provides food for the 7 billion people alive today and the more than 9 billion people by mid-century will in large part define whether a sustainable future is possible for humanity.

The models upon which food production has been predicated over the past century or so are unlikely to meet the challenges of the coming decades if the world is to overcome poverty and grow economies while also keeping humanity's footprint within ecological boundaries.

These challenges have been recognized for the last 20 years, including at the Rio Earth Summit of 1992 and underlined in a series of landmark reports including: the Millennium Ecosystem Assessment of 2005; the UNEP Global Environmental Outlook 4 and the Comprehensive Assessment of Water Management in Agriculture of 2007; the International Assessment of Agricultural Knowledge, Science and Technology for Development in 2009; and The Royal Society's 'Reaping the Benefits: Science and the Sustainable Intensification of Global Agriculture' report of 2010.

In *Living with the Trees of Life*, Roger Leakey, an agricultural research scientist with long field experience in the tropics analyses and presents case studies on how agroforestry offers innovative and compelling pathways towards food security, human well-being and environmental sustainability.

The book underlines how modern science and improved varieties of trees allied to centuries-old knowledge can provide a new set of marketable products of special importance to poor and marginalized people in the tropics and sub-tropics, while simultaneously rehabilitating degraded land and restoring soil fertility.

Unlike many books on the future of food and agriculture, this one does not fall into one or other camp in respect to the way forward. Dr Leakey draws on scientific and technical lessons from the Green Revolution of the 20th century, while also spotlighting those from agroforestry and organic and conservation agriculture.

Indeed the development of the argument in favour of multifunctional agriculture is a refreshing departure from the polarized and often sterile one-size-fits-all viewpoints that dog much of the food and agriculture debate.

The book deserves to be widely read: it is rich in imaginative but highly practical ideas – ones that offer real and tangible opportunities to transform subsistence agriculture while bringing degraded land back into production.

It offers a vision and a blueprint in which there is reduced pressure on forests; farmland with a more diverse set of crops and landscapes that protect watersheds are a better habitat for wildlife and sequester more carbon that, in turn, can help to combat climate change.

In doing so, it makes an important contribution to the transition towards the low carbon, resource efficient Green Economy so urgently needed in the 21st century.

Achim Steiner
UN Under-Secretary General
and Executive Director,
UN Environment Programme (UNEP)

Preface

> There are some common misconceptions about agroforestry that this book tries to address. Agroforestry is a low-input approach to agriculture, but it does not run counter to the Green Revolution. Instead, it is an approach to correct some of the mistakes of the Green Revolution and to increase the productivity of modern crop varieties. It thus aims to improve the returns on the investment in the Green Revolution. If widely adopted, it should then open new windows of opportunity for agri-business and help to achieve the origin objectives of the Green Revolution to overcome hunger, malnutrition and rural poverty.
> Roger Leakey (2012) The intensification of agroforestry by tree domestication for enhanced social and economic impact. In: *CAB Review: Perspectives in Agriculture, Veterinary Science, Nutrition and Natural Resources*. CAB International, Wallingford, UK.

To me the Garden of Eden conjures up images of lush vegetation in a land with plenty of natural resources where people live in equilibrium with their environment, eating alluring and little-known fruits and nuts – the 'Trees of Life'. I've lived and worked in the tropics most of my life and been lucky enough to see a few places where these images are not too far from the truth. In these places the Trees of Life are often as important today for food and balanced nutrition as they were hundreds and possibly thousands of years ago. These are not 'famine foods'; they are important resources for everyday life that are widely consumed and locally marketed. They have potential to be cultivated as new crops. Today, places where these species are found in profusion are the rare exceptions. Living and working in the tropics means that you far more often see places where these images are only a distant memory in the minds of the current older generation. The reality is that in many places Eden has been destroyed – forests are dwindling, the trees are becoming isolated, women are walking further to get fuel, people are hungry, soils are impoverished, the environment is damaged, erosion is causing landslides, unseen gases are escaping into the atmosphere and affecting the climate, and there is the smell of smoke from burning rubbish and vegetation, as well as from cooking fires. Make no mistake, the bad news, the 'doom and gloom' about tropical forests – over-exploitation of natural resources, the high incidence of poverty, malnutrition, hunger and disease – is real. These issues are not about to go away unless we find the political will to think very differently. Hopefully this book will open the eyes of people who can bring about this change in development policy; especially agricultural policy vis-à-vis poor smallholder farmers in the tropics.

Currently, despite decades of agricultural research, billions of people are still both poor and hungry and the planet is under threat from a food crisis arising from social deprivation and a range of environmental disasters that are associated with land degradation, and a changing climate. We are all familiar with such doom and gloom, but is there any hope of improving the situation? Based on our current experience, it seems that economic development and environmental protection are mutually exclusive; but is this really true? We have to ask ourselves: 'Have we genuinely tried to find a way to meet the food and wealth needs of all people in the growing population of our planet without causing environmental damage? Is economic growth achievable without over-exploitation of natural resources?' The answers to these questions have been elusive.

This book presents what I believe to be good news, as I have personally seen evidence from the tropics that convinces me that the world can easily support a population in excess of the currently projected 9 to 10 billion and that the poor can be substantially better off than they are now. I'm saying this as an agricultural scientist. I've been very fortunate to be personally involved in the work that has produced this evidence. So this book is an account of my personal experience, not just another report pulled out of a filing cabinet.

I will endeavour not to over-emphasize the good news and so give a false impression of the current situation. However, I hope that if enough people are encouraged by some good news it may be possible to hasten the development of a much-needed productive form of tropical agriculture that is also environmentally friendly. But what are the missing ingredients that make this dream so difficult to realize? Is it the loss of soil fertility or the lack of cash that prevents poor, smallholder farmers from harnessing the benefits of the 'Green Revolution'? Could solving these problems lead to a major new initiative in which food security and income generation run hand in hand with the rehabilitation of degraded farmland?

The prospect of good news may sound far too good to be true, but as a research scientist, I have spent much of my life trying to develop techniques and strategies that can be used by poor farmers in the tropics to empower themselves to greater self-sufficiency, and then to help them to build on that experience to raise themselves out of the 'poverty trap' that has ensnared them. To some extent this has meant the difficult task of trying to innovate in areas that are not fashionable in modern agricultural science, while also raising research questions that even now, despite the amazing things that are possible, are difficult to answer through science.

So is good news justified? Yes. It is also essential, as good news is usually drowned by bad news, which is more marketable than good news.

Without good news it is difficult to initiate and sustain any efforts to solve problems. Good news offers some reason for hope. In the absence of hope, despondency sets in and the only release for concerned people is to switch off, or to bury their concerns.

My story represents a journey, both geographically around the world to remote and interesting places in Africa, Asia, Latin America and Oceania, as well as through an unusual career in agroforestry,[1] developing practical solutions to the big global issues affecting mankind. Both journeys create worrying, amusing and interesting situations, some of which are woven into the more serious message of the book – how to overcome poverty, malnutrition, hunger, land degradation and even climate change, and so to recreate Eden. The Trees of Life are central to this story and have been very much part of my life. As we will see, I believe they can also play a critical role in the future of our world. So, I will try to weave these storylines together into the much bigger cloth of sustainable economic development with its over-arching social and environmental issues. Doing this as a biologist, I am aware that I will be skating on thin ice, as I am not a social scientist, economist or development expert. However, I have had the opportunity to work with people of many other disciplines and I hope some of their expertise has rubbed off on to me. Based on this assumption, I will make my case in an attempt to bridge some of the conceptual and professional 'disconnects' between disciplines and organizations, which I believe are part of the reason we have the big environmental and social problems facing agriculture worldwide. In a sense we have 'the blind leading the blind', as it seems that economists and policy makers often ignore the insights of biologists and ecologists, and vice versa – not a good recipe for economic growth with social and environmental sustainability.

It seems to me that another reason for some of our big global problems is that we – mankind – are not very good at examining the issues before suggesting solutions. I am deliberately excluding women from this criticism, as I think they have had a better vision of the issues, but sadly their voices have not been heard. I think agricultural development is a case in point. I therefore try to analyse the issues in Chapters 2 and 10, and then base the rest of the book on an approach that seems to me – another male – to address the pertinent issues.

I am aware that on occasions I will be stepping outside my area of expertise – my comfort zone. I will probably offend some of my colleagues and readers who are experts in these other fields. If so, I apologize. Nevertheless, I excuse my transgressions in the hope that by trying to set agroforestry in context, I will open a few eyes to new possibilities for the improvement of agriculture and Third World development. If you are an expert in one of the areas where I miss the mark, I hope that at least you will learn something about agroforestry and that you may be stimulated to improve the linkages between your work and that of agroforesters. In this way the number of disconnects can be reduced and the overall level of understanding among our peers be improved.

I will be bringing together information from my work about how we can encourage the spread of what I will call 'multifunctional farming systems', as well as how to improve their productivity and capacity to generate income. I will be explaining how agroforestry research in the tropics is 'domesticating' some of the Trees of Life by bringing them into cultivation as new crops that meet the needs of local people. None of this is rocket science. It is something that is readily attainable. However, success is dependent on politicians and policy makers having the will to think differently about how to reduce poverty, hunger and malnutrition. Actually, this new way of thinking is something we ought to be doing anyway, as it is also highly compatible with some of the lifestyle changes that we need to be implementing if we are to mitigate the impacts of climate change and conserve wildlife. So, if we put our minds to it, a better world, a new Eden, could be around the corner.

I've said that capturing the environmental, social and economic benefits of trees is not rocket science, and this is true. However, it does have complexity simply because it involves many different and interwoven forms of biological, ecological, environmental and social science. In this sense it is much more difficult than planting a field with a new variety of maize or rice. This complexity makes it relatively difficult to explain over a drink in a bar or at a dinner table, when someone asks 'and what do you do?'. I know that I have failed to get the message across on many such occasions, hence my decision to try to get it all down in black and white in this book, in what I hope is an understandable manner.

'What do you do?' is a question that, like most people, I am frequently asked in everyday conversation. The questioner is expecting a simple answer: 'oh I'm a doctor/solicitor/accountant/businessman'. But for me the answer is not that easy. There are dozens of potential answers that I can give to this question, so my reply tends to reflect my mood, or the sort of answer that I guess might interest the questioner. So, what are the answers? Well, there are those that reflect my employer, things like 'well, I'm a civil servant' or 'oh, I'm an academic', 'I work in overseas development', 'I'm a consultant'. Alternatively, there are the academic disciplines that I am engaged in: 'I'm a forester/horticulturalist/agriculturalist/ecologist/agroforester/food scientist/tree physiologist'. Once, at a party, the last of these options elicited a blank look, the questioner moved on to other chit chat with people standing nearby, and then came back half an hour later: 'Tell me, what does a tree psychologist do?' If I had not been so taken aback by this question, I might have answered 'Oh, well, the buzz of a chain saw gives trees the jitters, you know. We talk to trees and try to sort out their emotional problems – these typically stem from the rate of deforestation.'

Disciplines and professions are not really a very good answer to the question 'what do you do?', as I really work at the interface between all these disciplines, and I could perhaps add 'social science'. Therefore it is sometimes easier to describe the ways that I spend my time: 'I'm a research scientist/writer/fund raiser/manager/administrator,' or 'I'm in agribusiness'. If I'm feeling flippant, I could add 'I travel a lot, visiting airports all around the world and see how well they are functioning'. If, on the other hand, I feel that the person I am talking to is genuinely interested to know what I do, and not just making conversation, then I can answer: 'Well, I am developing techniques and strategies to try to overcome some of the big problems facing people in developing countries around the tropics – things like poverty, deforestation and environmental degradation, and malnutrition.'

Now we are getting to the heart of the matter. So, *Living with the Trees of Life* is my attempt to answer the questions 'what do you do?' and 'why do you do it?' more efficiently than I have ever managed in any conversation, with anyone, including, I suspect, my wife and family, and certainly many friends and work colleagues. Many friends and acquaintances are just aware that I am always coming back from, or about to go to, some remote corner of the world. Many of these people, especially some of my former bosses, think that my life is one long holiday travelling to exotic locations. On many, many occasions this is far from the truth and I am staying in remote villages, eating unusual local food, sweating and swatting mosquitoes or the vectors of other nasty tropical diseases. What makes this worthwhile is that my experience of farmers in developing countries is the friendliness of their welcome and their willingness to interact and participate in studies involving their farms, despite their hardships, poverty and other problems. I also spend a lot of time milling around in boring places like airports and in the offices of government officials waiting to get permissions and agreements. My waiting skills are well honed.

I hope that through my experience I have some insights into the complex multidisciplinary issues surrounding the sustainability of agriculture in the tropics. One of the problems of modern life is that we are all trained in the ever increasing detail of our different disciplines. As a consequence we are less well equipped when we get into jobs in which interaction between disciplines is needed to address the numerous disconnects in the way we view issues of everyday life. Indeed, it seems to me that the world now needs people who are trained in 'multidisciplinary studies'. I hope this book will give people from very different disciplines an insight into a more multidisciplinary approach to agriculture and to the resolution of some of the big global issues that stem from our current approach.

Finally, as a biological scientist, I am used to justifying my remarks by reference to the scientific literature. In this book I have kept this to a minimum in an attempt to make the text more readable. This could leave me open to criticism by 'experts'. To reconcile this predicament, I present some of my formal publications in 'Further Reading' for those who wish to assure themselves that my remarks are supported by academic literature.

Professor R.R.B. Leakey

rogerleakey@btinternet.com

Note

[1] Agroforestry is an umbrella term that covers a wide range of land use systems in which trees are integrated into the farming system.

Acknowledgements

I especially thank my parents for the life-shaping experiences of my childhood and the sacrifices they made for my education. My career and family life started at the same time, with my marriage to Alison Swan a few months after landing my first job at the Institute of Tree Biology. This book is about my career and working life, and not the family. However, my career, especially the overseas travel, interrupted family life. I was away from home about half of the time, usually on trips of 1–3 weeks, although occasionally longer. This meant, of course, that much of the daily burden of household jobs and looking after the needs of our boys, Andrew and Chris, fell on my wife Alison. She suspended her teaching career for many years to take on extra duties. My travels also meant that our boys grew up with a father who was unable to fully fulfil his parental duties. I greatly regretted this, but just occasionally I could provide some compensation by hooking a family holiday on to an overseas trip. I'm pleased to say that my absences do not seem to have had serious detrimental effects on the boys, who have grown up and done extremely well. They are now in the early years of their own careers, also in biology and environmental sciences. I am very grateful to Alison for making such a good job of being mother and part-time father.

I also take this opportunity to sincerely acknowledge everyone who has contributed to the richness of my career experience, especially my numerous work colleagues who have contributed to my knowledge and the formulation of my ideas. I also thank the numerous donors who funded our work, as well as the community partners and farmers in many parts of the world who made it all happen. In the chapters that follow and the endnotes, I mention many of you, but I certainly have not mentioned all of you – so please don't take offence if you have not been mentioned. It doesn't mean that your work was any less appreciated; it is just that there is a limit to what I can include in this book.

I am very grateful to the following, who have read and commented on my manuscript: Ian Donaldson, Joan Baxter, Dick Chancellor, Craig Elevitch, Alison Leakey, Chris Leakey and Andrew Leakey. In addition, I gratefully acknowledge the help of Claire Parfitt, Tracy Head and Maggie Hanbury in the publication of this book.

Permissions

Permissions to reprint have been granted by Nova Publications (Fig. 2.1), Commonwealth Science Council (Fig. 7.4) and the *International Journal of Agricultural Sustainability* (Table 9.1). Figure 9.2 was provided by Daimler AG with permission to reproduce.

Photographs

With the exception of Fig. 9.2, the photographs were taken by the author.

Frequently Used Acronyms

ACIAR	Australian Centre for International Agricultural Research
AFTP	Agroforestry tree product
CATIE	Centro Agronómico Tropical de Investigación y Enseñanza in Costa Rica
CEH	Centre for Ecology and Hydrology (formerly Institute of Terrestrial Ecology, ITE) of Natural Environment Research Council (NERC) in the UK
CGIAR	Consultative Group on International Agricultural Research
DFID	Department for International Development (formerly ODA, see below) in the UK
FAO	Food and Agriculture Organization of the United Nations
IAASTD	International Assessment of Agricultural Knowledge, Science and Technology for Development
ICRAF	International Centre for Research in Agroforestry (now the World Agroforestry Centre) in Kenya
JCU	James Cook University, Cairns, Australia
ODA	Overseas Development Administration (now DFID) in the UK
UNESCO	United Nations Educational, Scientific and Cultural Organization

Revelations in Kumba 1

Each generation of forest trees typically arises from virtually wild stock that has not even had the centuries of conscious selection that successive small-holders have given to many food crops. Would it not be prudent to devote substantial sums to research on the genetics and domestication of important tree genera? After all, the returns should theoretically be considerably greater from such unimproved species than from those which must now be approaching their maximum potential.
>   Alan Longman and Jan Jeník (1987) *Tropical Forest and its Environment*,
>   2nd edn. Longman Scientific and Technical, Harlow, UK.

Trees influence landscape scale dynamics more than any other organisms. Investigation of this keystone role must remain at the very heart of the research agenda because of the huge number of secondary interactions that flow from the incorporation of trees within any land use system.
>   Mike Swift *et al.* (2006) Confronting land degradation in Africa: challenges
>   for the next decade. In: Dennis Garrity *et al.* (eds) *World Agroforestry
>   into the Future*. World Agroforestry Centre, Nairobi, Kenya.

I'm in Kumba market, in the South-west Province of Cameroon, looking at stalls laid out with a wide range of unusual looking fruits, nuts, dried tree bark and other products that I cannot identify (Fig. 1.1), but I am aware that they are the products of the Trees of Life. I'm with Patrick Shiembo, a student who has recently registered to do a PhD at Edinburgh University. I did not know it at the time but I was on the brink of a career change. For years I had been a research scientist at the Institute of Tree Biology,[1] just outside Edinburgh. My work involved the development of techniques for the domestication of some of West and Central Africa's most important local timber species; especially one known as 'obeche' in Nigeria, or *Triplochiton scleroxylon* to give it its scientific name. Like many of the important tropical hardwoods it had seed supply problems, so among other things I had been developing techniques to propagate the species

Fig. 1.1. A market stall in Kumba, Cameroon, selling a range of indigenous fruits, nuts and other products from indigenous trees.

vegetatively using stem cuttings. Patrick's PhD project was going to see if these techniques could be adapted to three local tree species that were all important foods – well, actually one was a woody vine. We were in Kumba market so that Patrick could show me what these species looked like.

The first is called the bush mango. It is no relation of the common mango, which comes from Asia originally, although now it is grown all around the tropics and its fruits are in supermarkets all around the world. The bush mango (*Irvingia gabonensis*) is a similar shape, size and colour to the common mango. However the flesh of this wild African fruit is both stringy and somewhat bitter and astringent – in other words, not very tasty unless you are hungry. Like a mango it has a big hard seed and if you manage to crack the nut without damaging your fingers, there is a kernel that looks not too unlike a butter bean (Fig. 1.2). It is this kernel that is the

Fig. 1.2. Bush mango (*Irvingia gabonensis*) kernels for food thickening.

important food. When it is boiled it exudes a slimy substance that is highly appreciated by local people as a food thickening agent, and so is a common ingredient in soups and stews. These kernels are widely traded and a good source of income, especially for the women who generally are the stallholders in local markets. The importance of such income was made clear to me one time when I was visiting the Minister of Agriculture in Cameroon, Dr Ayuk Takem, in his office. He informed me that his mother had sold bush mango kernels and that it was this income that had allowed him to be educated.

The second species is njangsang (*Ricinodendron heudelotii*), sometimes called the peanut tree. This green fruit is about the size of a golf ball, and it too contains an edible, pea-sized kernel, but this time it is used as a flavouring, or spice, often with fish dishes (Fig. 1.3). Again, piles of these seeds are widely traded in local markets.

The third species, not seen in Kumba market the day that I was there, was eru (*Gnetum africanum*). It is a very nutritious leafy vegetable, rich in protein, essential amino acids and minerals, especially iron, that is eaten as a spinach (Fig. 1.4), often well spiced. Eru is interesting as it is a vine that grows in deep shade and so has the potential to be grown under other important tree crops. Botanically it is also interesting as it is related to conifers, although superficially it looks just like any broadleaved plant. It is its cone-like fruits that give away its ancient and primitive ancestry. Freshly chopped leaves of eru are usually sold in small piles in local markets, but whole leaves are also exported from Cameroon to Nigeria in grossly overloaded vans. This trade is severely depleting the natural resource. In fact eru is becoming rare in the forest. This is serious for such an important source of nutrition.

Bush mango and njangsang are increasingly harvested from trees growing in areas of farmland, where natural seedlings are protected by farmers when they till the land. Eru, on the other hand, is still harvested almost entirely from the shade of remnants of moist tropical forest. It is

Fig. 1.3. Kernels of njangsang (*Ricinodendron heudelottii*) in Cameroon.

Fig. 1.4. Eru (*Gnetum africanum*) leaf in a market in Yaoundé, Cameroon.

hard to grow in farmland as methods to germinate the seeds are unknown. This made Patrick's work to develop an alternative means of propagation extra important. Fortunately, Patrick found that eru was easily propagated by cuttings and so now there is the possibility of growing it in a shady corner of a garden. More recently, Joseph Nkefor of Limbe Botanic Gardens in Buea, on the coast of Cameroon, has been studying its light requirements for rapid growth. Thus between them, Patrick and Joseph have overcome the practical problems threatening the survival of this species and now it is increasingly being successfully cultivated in the villages.

Somehow, throughout my life, it seems that I have been continually drawn towards an interest in biology and the environment; and a desire to work with, learn from and hopefully improve the lot of poor farmers, especially in Africa. Yet, as we will see, many of the conscious decisions that I have taken have not really been headed in those directions. So it seems there was an element of destiny in the direction that my life and career took. To some extent, no doubt, my love for Africa can be attributed to the fact that I was born and brought up in Kenya, surrounded by its beautiful scenery and wonderful wildlife. My parents certainly encouraged my interest in wildlife. In my school holidays I made frequent trips with my parents to Nairobi Arboretum, the City Park and local forests to catch butterflies and then to see Dr Robert Carcasson, the entomologist at the Coryndon Museum (now the National Museums of Kenya). He taught me how to keep records of the butterflies that I had collected, and helped me to name them.

Maybe with this childhood, it was inevitable that I would want a career that didn't confine me to an office, and that tropical forests would hold a special attraction. However, a career was not something on my mind at that time and indeed schooling was not something in which I excelled, even when I went on to secondary school in England. That is, until I started to do biology in the fifth form. Biology was great. Here I had found my academic niche. It was just so interesting that I didn't even have to try to learn it. Everything I read about just seemed to stick in my brain.

All too soon, however, it was time to think seriously about a career. Although my father and grandfather had been foresters in Kenya and India, respectively, I didn't choose forestry. I decided to train to be an agriculturalist. Initially, I thought that I would do farm management, and so I registered to do a practical agriculture course at Seale-Hayne Agricultural College in South West England, just outside the Devonshire market town of Newton Abbot. Before doing the course work, all students had to have 18 months' practical experience. So I worked as a farm student/labourer on a dairy and arable farm near Blandford in the county of Dorset, before going to Sweden to work on a farm in Ljustorp, north of Stockholm, halfway up to the Arctic Circle. Then, finally, I worked on a sheep farm in the Highlands of Scotland, beside Loch Rannoch. Two years later I graduated from Seale-Hayne with a National Diploma, but by this time I had decided that I would never be able to afford a farm of my own and that I didn't want to manage other people's farms. So, I decided to do a degree at Bangor University in North Wales.

When I registered with the university, I hoped to include fish farming in my agriculture course, but this subsequently turned out to be impossible, so I decided to read Agricultural Botany, as I had been greatly impressed by the Plant Biology Department. As things turned out, this decision was a pivotal moment in my career, as it was this course that triggered my interest in research and in finding out what makes plants 'tick'. I found the combination of plant physiology, genetics and ecology very interesting, and as we will see they provided a foundation for what I was to do in later life.

Of course student life is not all work and one evening in a local pub with some friends[2] we decided to do an overland expedition to India. Most of us were 'agrics', so we decided to go to northern India, where the Green Revolution was underway to address the global problem of hunger. Institutes and universities in India were at the forefront of this huge agricultural research initiative, so we made plans to visit these places and to see the Green Revolution first-hand for ourselves. To help us with fund-raising and to raise our profile, we initiated a university Exploration Society and spent much of the second year at university seeking, very successfully, sponsorship from industry. In June 1969 we set off on the first leg to Istanbul. Then after a little sightseeing our second leg took us across Turkey via Ankara and then the Black Sea coast to Trabzon and then we turned inland climbing up through a splendid mountain pass to Erzurum, before passing the snow-capped peak of Mount Ararat – there was no sign of Noah or his Ark – and entering Iran and driving to Tabriz and Tehran.

After Tehran, we headed for Afghanistan and moved into the desert and one evening looked up at the moon just as Neil Armstrong and Buzz Aldrin made their historic landing. In those days crossing Afghanistan was relatively easy, although there were stories of bandits. After the Khyber Pass we crossed Pakistan and arrived in India 3 weeks after leaving England. Our plan was to visit a number of agricultural universities and other projects to see for ourselves how agricultural research was tackling the issues of hunger and poverty. Our first stop was at the new Punjab Agricultural University, Ludhiana (now Guru Angad Dev Veterinary and Animal Science University), where we spent several days learning about the Green Revolution's battle to raise food production through a combination of crop breeding and the efficient use of inputs such as fertilizers and pesticides. It was here that I first heard about the challenge to disseminate new technologies to farmers. The farmers we visited were keen to test the new crop varieties as the early results were very encouraging, however their excitement was tempered with some doubts as to their ability to achieve the same levels of success by themselves.

From Ludhiana we headed for Delhi and then after a few days we continued east into the State of Uttar Pradesh before turning north up towards the Himalayas. Here we visited Uttar Pradesh Agricultural University in Pantnagar (now the GB Pant University of Agriculture and Technology), described by Nobel Laureate Norman Borlaug as 'the harbinger of the Green Revolution'. There we were taken out to see the Green Revolution research plots where new crop varieties were being tested with and without the recommended applications of fertilizers. The best combinations were producing grain yields 300–400% above the average achieved by local farmers. We also visited some small farms in the area to see how these technologies were being taken out into farmers' fields through the university extension services, to promote their adoption by local farmers. It was evident from this that the farmers found the enormous increases in grain yield both exciting and somewhat daunting. Obviously the improved yields meant that hunger would be alleviated and that these farmers would have adequate food for their families, but as we had also heard at the Punjab Agricultural University, the farmers were unsure that they would be able to replicate the results without the continued input of university staff and extension workers. One farmer voiced these concerns by saying: 'We don't know if this magic will work for us'.

Having achieved the second of our four agricultural visits we drove the 65 km north towards the foothills of the Himalayas and towards the hill station of Nainital (1934 m), where my mother had been born in 1907. My grandfather had been stationed there in the Forestry Department before he returned to England to teach forestry at Cambridge University.

Our next destination was Patna in Bihar State, close to the River Ganges. The state had been the centre of the much publicized famine of 1966–1967. The coincidence of the famine with the early successes of the Green Revolution in India did much to raise the profile of the international effort to address global famine. Our plan was to visit an Oxfam project that was following up on some of Mahatma Gandhi's ideas for social reform. We spent 2 days visiting villages to see the ways in which a peasant form of agriculture was being enriched by wider participation, use of appropriate technologies like biogas, and improved education and health services.

From Bihar we turned back towards home visiting a few Hindu and Sikh temples and the Taj Mahal along the way back to Delhi. From there we went to Quetta in Pakistan to visit the large cotton plantations of Sir William Roberts, an eminent agriculturalist in the Punjab, who was also an 'alumnus' of Bangor University, and who had agreed to sponsor our expedition if we visited his estates. The contrast between these and the smallholder farms that we had seen in India was very pronounced and certainly gave us the chance to consider the relative merits of the different livelihood options we had seen. Although at first sight the two extremes, epitomized by a struggling peasant farmer and a farm labourer employed by a large plantation company, might seem to be balanced in favour of the latter. I think in reality most people would rather be independent. The challenge for agriculture is to find a way to improve the lot of smallholder farmers and their families. This means enhancing the livelihood benefits flowing from farming and includes some of the ideals expressed by Gandhi, things like dignity, equity, culture, self-sufficiency, health and education.

Our final stop before heading for home was to visit an Iranian friend who had a watermelon farm that he had created out of the desert by installing irrigation. It was very educational to see first-hand the huge impact of irrigation on crop production in an area like this. I hope these benefits persisted, as there can be risks of salinization when the salts dissolved in well water get deposited at the soil surface. Certainly there was no sign of this at the time of our visit, which by pure luck coincided with harvest time and so after driving around the farm in his jeep, we sat in a field in the full heat of the day and ate ultra-fresh watermelons. They were very refreshing.

Amazingly, we got back to England on the very day we had told our parents we would. This trip had certainly opened my eyes to the achievements of the Green Revolution, as well as to some of the issues facing farmers and poor communities in developing countries. Like the practical experience that I had as a farm student in the UK and in Sweden, this experience had provided a useful foundation on which to build my later career. It also whetted my appetite for travel.

When I graduated from Bangor, I had a choice of going out to find a job or, as my professors encouraged me to, go on with further education and do a PhD. The vacancies advertised at that time were mostly in agricultural chemical companies. I had several interviews and was asked, 'how do you see your career developing and what do you really want to do with your life?'. I remember saying that I wanted to help poor people in developing countries to produce the food they needed. I never thought that I would get a chance to do something like that. None of the jobs really appealed to me and once again things fell into place, and I got the chance to do a PhD registered at Reading University but actually working at the Weed Research Organization in Kidlington, just north of Oxford. Three years later as a qualified research scientist, I could no longer postpone my entry into employment. Luckily, I was soon offered the chance to join the Institute of Tree Biology (ITB) at the Bush Estate, between Edinburgh and Penicuik, on a 3-year contract from the Overseas Development Administration, as part of a British Government Aid package to Nigeria.

My project in ITB was to provide technical support to the Forest Research Institute of Nigeria (FRIN) in Ibadan by developing robust tree propagation techniques for the indigenous hardwood tree obeche (*T. scleroxylon*).

As seed was not readily available, the idea was to use stem cuttings (a piece of stem with a leaf – Fig. 1.5) to mass produce planting stock for a forest regeneration programme. At this time many people believed that it was virtually impossible to propagate tropical timber trees in this way. Worldwide only a handful of timber tree species were considered amenable to this approach – poplars and willows in Europe and North America and *Cryptomeria japonica* in the Far East. However, good progress was being made at about this time in a few *Eucalyptus* species in the Congo.

My work with obeche progressed well, and my contracts were renewed and over a period of about 20 years my colleagues and I had made sufficient progress to propose some general principles that we believed applied to most tree species. We will consider these principles in some detail in a later chapter (Chapter 7). Over these 20 years we also expanded our forestry activities to a range of other timber trees from Africa, Asia and Latin America. The consequence of these forestry projects was that we had gained a great deal of knowledge about the biology of tropical trees, especially with regard to vegetative propagation and clonal forestry

Fig. 1.5. Rooted cutting of a single-node stem cutting of *Triplochiton scleroxylon* (obeche).

(see Appendix for details), and had acquired experience of the countries in which we worked. This led in particular to the recognition of our work on the techniques and strategies for the domestication of tropical trees and is how I came to be with Patrick in Kumba thinking about his work to propagate bush mango, njangsang and eru.

It was a day or two after the visit to Kumba market that the significance of seeing the diverse collection of fruits, nuts, leaves, barks and resins in Kumba market really hit me. These products are sometimes called 'famine foods', but this is a serious misnomer. They are much more important than that name implies. These are the products that, as hunter-gatherers, local people used to harvest free from the forests and woodlands and now they are commodities that are marketed locally and regionally, and not just in trivial quantities. Products derived from some of these species are breaking into international markets. This change from wild food to marketable product is the result of forest clearance for agriculture, roads, towns and many other trappings of modern life. Extraordinarily, I had not put two and two together before. When you are trained to think in a narrow way particular to one discipline, it takes a revelation to open your eyes to another way of more holistic thinking. Of course I knew that tropical deforestation was causing serious environmental damage – the newspapers and other media have been saying so for years – but I had always assumed that the social consequences of 'development' were beneficial, with the pros outweighing the cons. Suddenly, there were question marks written all over this way of thinking. These are not the 'famine foods' they are often portrayed to be; they are really popular and useful additions to the diet of local people, offering new opportunities for agriculture. Development is depriving local people of highly nutritious and traditionally important foods – we could be cultivating them, and recreating Eden.

After the revelation in Kumba, the seed of an idea already in my head suddenly germinated. The tree domestication package that we had developed for timber trees in Edinburgh had enormous potential to improve and domesticate the numerous and traditionally important tree species producing foods, medicines and other marketable products useful to local people throughout the tropics. Moreover, the cultivation of these trees would also enrich and diversify agricultural systems. To bring this germ of an idea out into the scientific arena, Adrian Newton and I organized and hosted an international conference, entitled 'Tropical Trees: the Potential for Domestication and the Rebuilding of Forest Resources'. It was this germ that was to grow and subsume my life and open up doors for a new phase of my career – back in agriculture.

At this conference in Edinburgh we coined the term 'Cinderella' trees for all those invaluable species that had basically been ignored by formal science, thus aligning them with the children's story in which the dominant, fun-loving ugly sisters were blind to the beauty and talents of the hard-working Cinderella until she was swept away to marry Prince Charming. This came about thanks to the kindness of her fairy godmother and her previously hidden charms. In Kumba I had realized that the huge array of traditional foods also had hidden charms; what we needed was a fairy godmother. At the time of the conference we foresaw the domestication of new tree crops as the entrée into a 'woody plant revolution', to match the Green Revolution in agricultural crops. It still has some way to go to achieve that level of significance, but in the 18 years since that meeting things have certainly moved in that direction.

Looking back now, this Domestication conference was the start of new era in which the emphasis of our research work in Edinburgh shifted from forestry towards agroforestry – the integration of trees in farming systems. This change came about quite quickly, as by coincidence the Department for International Development (DFID – the successor to ODA, the Overseas Development Administration) developed a new research strategy for UK development aid – this was more socially oriented. Thus the benefits from research had to flow back to rural communities, rather than to forestry companies or the national exchequer. As a result it was becoming increasingly difficult to obtain research funds for our work relating to plantation forestry.

A short while after the Domestication conference in Edinburgh, I applied for the position of Research Director at the International Centre for Research in Agroforestry (ICRAF), headquartered in Nairobi.[3] I based my application on the need for agroforestry to adopt the principles of domestication and to apply them to the wide range of species that farmers in the tropics used to gather from natural forests and woodlands. This idea was accepted by ICRAF management and I was soon sitting in my new office, with a new career as an agroforester ahead of me.

Now, I had to determine how agroforestry could contribute to modern agriculture and improve the lives of smallholder farmers in the tropics. My gut feeling at this time was that it had something to do with the traditionally important trees producing products traditionally gathered from natural forests. This was to be the legacy from my visit to Kumba market.

Notes

[1] Institute of Tree Biology (ITB) later changed its name to Institute of Terrestrial Ecology (ITE) and then following further restructuring to Centre for Ecology and Hydrology (CEH).
[2] Dick Esslemont, Simon Stubbs, Nick Shapland, Bill Forse and Roy Jelly. In the end Dick and Bill were unable to join the expedition.

³ ICRAF was conceived in 1975 and became a reality in 1977. It then joined the Consultative Group on International Agricultural Research (CGIAR) as one of the elite International Agricultural Research Centres in 1991 and opened regional offices in southern Africa, humid West and Central Africa, Sahelian West Africa, Latin America and South-east Asia. In 2002, it changed its name to World Agroforestry Centre.

Further Reading

Leakey, R.R.B. (1990) The domestication of tropical forest trees by cloning: a strategy for increased production and for conservation. In: Weiner, D. and Muller, P. (eds) *Fast Growing Trees and Nitrogen Fixing Trees.* Gustav Fischer Verlag, Stuttgart, Germany, pp. 22–31.

Leakey, R.R.B. and Newton, A.C. (1994a) *Tropical Trees: the Potential for Domestication and the Rebuilding of Forest Resources.* HMSO, London.

Leakey, R.R.B. and Newton, A.C. (1994b) *Domestication of Tropical Trees for Timber and Non-timber Forest Products.* MAB (Man and Biosphere) Digest No. 17, United Nations Educational, Scientific and Cultural Organization (UNESCO), Paris.

Leakey, R.R.B., Last, F.T. and Longman, K.A. (1982) Domestication of forest trees: a process to secure the productivity and future diversity of tropical ecosystems. *Commonwealth Forestry Review* 61, 33–42.

Longman, K.A. and Leakey, R.R.B. (1995) La domestication du Samba. *Annales des Sciences Forestières* 52, 43–56.

Shiembo, P.N., Newton, A.C. and Leakey, R.R.B. (1996a) Vegetative propagation of *Irvingia gabonensis* Baill., a West African fruit tree. *Forest Ecology and Management* 87, 185–192.

Shiembo, P.N., Newton, A.C. and Leakey, R.R.B. (1996b) Vegetative propagation of *Gnetum africanum* Welw., a leafy vegetable from West Africa. *Journal of Horticultural Science* 71, 149–155.

Shiembo, P.N., Newton, A.C. and Leakey, R.R.B. (1997) Vegetative propagation of *Ricinodendron heudelotii* (Baill) Pierre ex Pax, a West African fruit tree. *Journal of Tropical Forest Science* 9, 514–525.

The Big Global Issues 2

> To sum it up, the challenge facing the world's 1.8 billion men and women who grow our food is to double their output of food – using far less water, less land, less energy and less fertilizer. They must accomplish this on low and uncertain returns, with less new technology available, amid more red tape, economic disincentives, and corrupted markets, and in the teeth of spreading drought.
> Julian Cribb (2010) *The Coming Famine: the Global Food Crisis and What We Can Do to Avoid It*. University of California Press, Los Angeles, California.

> The Green Revolution has won a temporary success in man's war against hunger and deprivation; it has given man a breathing space. If fully implemented, the revolution can provide sufficient food for sustenance during the next three decades.
> Norman Borlaug (1970) *The Green Revolution, Peace and Humanity*. Nobel Peace Prize lecture, Oslo, Norway, 11 December 1970.

In the dim and distant past most people in tropical countries had access to some areas of forest where, as in my mental image of the Garden of Eden, they could gather a wide range of fruits, nuts and other useful products for food, medicines and all the other necessities of life. Things started to change with the arrival of colonists from Europe, who saw the wealth of the natural resources of the tropics. Very soon these riches became the source of raw materials for overseas industries and of commodities for daily domestic consumption in Europe. While colonization spread and intensified, the expansion of industry started a chain of events, which brought prosperity to the industrial nations and started the process we now call 'globalization'.

In general, early 'colonists' were not altruistic in their intentions, seeing an opportunity to exploit the natural resources of the tropics either for 'Queen and Country' or for their own gain, or both. At first, timber and spices were harvested from the wild, but later large-scale plantations were established for commodities like cotton, sugar, rubber, tea and coffee for export to industrialized countries. With this came the subjugation of local people as labourers. Back at home, however, industrialization created business opportunities and employment outside agriculture and a rise in living standards. Similar development did not occur in the tropical countries. Thus today in developing countries, 80% of people in the rural population are farmers, while in industrialized countries the equivalent figure is around 1–3%.

Greater affluence in the 'West' is, however, offset to some extent by a higher cost of living, although when times are hard, many people in industrial countries can get social and financial support from government resources. In contrast, in the tropics there is little or no social support and so when times are hard people have to rely on their extended family for support. This means that farming families have to be self-sufficient for food and all the other everyday needs of life.

With colonialism and its associated economic and social changes, traditional ways of life in the tropics were swept aside, eroded and replaced with new and often foreign ideas influenced by the Industrial, Chemical and Mechanical Revolutions of 18th, 19th and 20th centuries. In her book *The Challenge for Africa*,[1] Nobel Peace Laureate Wangari Maathai has forcefully explained how colonial life led to a loss of traditional culture, collective power and social responsibility, which she says makes it difficult for Africans to cope with the pressures and ways of modern life. Part of the colonial ethic was to discourage the use of traditional local foods from the forest, in the name of 'civilization', and to replace them with foreign food crops for cultivation on land cleared of woody vegetation. This was the start of the philosophy of modern agriculture in the tropics. It was after my visit to Kumba market that I started to raise question marks in my mind about the wisdom of this philosophy. Since that visit, many African farmers have told me how much they would like to cultivate the traditionally and culturally important species in their farms.

Despite the negative impacts of colonialism and globalization and their associated social and economic 'advances', the world population has exploded – doubling over the last 40 years. This population growth is projected to stabilize at about 9–12 billion towards the end of this century. Of course, more people means the need for more food. Thus it seems certain that we can expect an ever-increasing strain to be placed on agriculture, the land and the environment. So far, three ways to alleviate this strain have been tried. The first was implemented by shifting cultivators. When they ran out of forest they just cleared an area of fallow before it had fully returned to forest and replanted it, so shortening the fallow period. This prevented the full replenishment of soil fertility by the new fallow vegetation and, as a result, both the nutrient and the carbon stocks in the soil declined. This has resulted in widespread land degradation and declining crop productivity.

Secondly, either as an alternative to, or as a consequence of the first approach, farmers have moved to new areas of forest, often in the wake of logging companies, and opened up new farms and new communities. It is this which has drastically reduced the area of tropical forest still standing.

Thirdly, under instruction from government extension agents, farmers made permanent settlements and became sedentary farmers cultivating the same area of land over and over again using the new Green Revolution package of high yielding varieties and artificial inputs of fertilizers, pesticides and, when necessary, irrigation. As I saw during my student travels in India in 1969, this third option can allow massive increases in agricultural production. These gains have been realized in some places over the last 100 years, but not everywhere – and especially not in Africa. Despite these gains the strain on production has not gone away. Partly, this is due to the ever-increasing population and the reduction in farm size that occurs as parents' land is partitioned among their children. Partly, and perhaps more importantly, huge numbers of farmers in developing countries have lived, and in many cases still live, at a level close to subsistence. Often these farmers are living virtually outside the cash economy, locked in poverty, and they are unable to afford the fertilizer and other inputs that make the Green Revolution package viable. As a result, the farmland becomes so infertile that the yields of crucial staple food crops go into decline. They are then also at risk of falling into malnutrition and hunger. One dreadful statistic that illustrates this is that 75% of smallholder farmers in the tropics are malnourished; and these are the people who are the food producers!

Let's look at the Green Revolution in a bit more detail. It was, and still is, a massively ambitious programme of plant and animal breeding aimed at the production of enough food to eliminate hunger. The prime focus was on breeding programmes to increase the yield potential and quality of staple food crops such as rice, maize, wheat, cassava and other root crops. In parallel with this, new agricultural technologies such as irrigation, farm mechanization, artificial fertilizers and pesticides were developed to intensify the production when grown in monocultures. In many ways the achievements of the Green Revolution's programme of agricultural research over the last 60 years have been a dramatic success. Indeed, the level of success has probably far exceeded the hopes of the early 'fathers' of the programme.

Global agricultural production today is many orders of magnitude greater than it was 50–60 years ago and now supports a world population of 7 billion people. This is attributable to modern agriculture in many parts of the world being highly intensive and mechanized and supported by a huge multi-billion dollar agricultural industry (over US$12 billion for research and development in 2000). On the downside, however, the economic and production benefits of modern agriculture have not been equally distributed around the world. The countries of the temperate zone have been by far the greatest beneficiaries of the technology, as large-scale, mechanized monocultures are consistent with the greater wealth of the farming community in industrial economies and the small proportion of their overall population engaged in farming. In addition, the industrial economies were in a favoured position to develop the agricultural industries manufacturing fertilizers, pesticides and farm machinery, as well as having the research capacity to engage in crop and livestock breeding. On top of all that, these countries already had the infrastructure and experience of several hundred years of worldwide trade.

The situation in Africa and some countries of Asia and Latin America was very different. So it is not surprising that the benefits of the Green Revolution have been lower here. Nevertheless, overall production in Asia and Latin America has greatly increased with, for example, a doubling of average cereal yields. For the people of Africa, however, the net production benefit has been a modest 12%, and much of this has come at the expense of a large expansion of the area under production. This dilemma should not really be a surprise to us. Dr Norman Borlaug, who played a leading role in the 'Green Revolution', foresaw that it was not going to be a permanent solution. At his investiture in 1970 as a Nobel Peace Prize Laureate, he said that the 'Green Revolution' has only 'won a temporary success in man's war against hunger and deprivation; it has given man a breathing space'.

It seems that this breathing space has nearly expired and we are again facing a food crisis. The complexities of the reasons for this are eloquently explained by Julian Cribb[2] in his book *The Coming Famine: the Global Food Crisis and What We Can Do To Avoid It*. The causes boil down to greed, the excessive overuse of natural resources, poor governance and social conflict. He offers many practical solutions to solving the crisis. They all involve us taking greater responsibility over how we manage natural resources, and how we manage the food industry. Julian sums up the challenges posed by the global food crisis as: 'to double the output of food – using far less water, less land, less energy, and less fertilizer … amid more red tape, economic disincentives, and corrupted markets, and in the teeth of spreading drought'. That alone is an enormous challenge, especially when we still have half the world burdened with hunger, malnutrition and poverty. It's clear from current statistics that we are a long way from solving the problems (Table 2.1).

Globally, about 0.9 billion people are still suffering from hunger due to lack of calories and protein, while 2 billion people are suffering from mineral and vitamin deficiency as a result of a lack of nutritious food. Sadly, many of those who are hungry and undernourished are also among the 3.2 billion people (47% of world population) in poverty with an income of less than US$2/day. Over 1 billion people still live on less than US$1/day. We must, however, remember that agriculture is not the only cause of these statistics; wars and natural disasters also contribute to the misery of

Table 2.1. The scale of the big environmental and social problems.

Problem	Impact	Scale
Environmental and natural resource problems		
Land degradation	Loss of soil fertility	2 billion ha (38% of the world's crop land)
Depletion of water resources	Loss for agricultural uses	2664 km^3/year (70% of global freshwater)
Depletion of soil nutrients	N, P, K deficiencies	59, 85 and 90% of harvested area, respectively
Salinization	Rising salt due to evaporation from soils	34 million ha (10% of irrigated area)
Deforestation and loss of vegetation	Loss of biodiversity and agroecological functions	Valued at US$1542 billion/year
Increasing pollution (including pesticide and fertilizer)	Reduced water quality in rivers	1.5 billion people lack safe drinking water
Increased greenhouse gas (GHG) emissions	Climate change	Agriculture responsible for 15% of GHG emissions
Social and economic issues		
Poverty	Income less than US$2/day	3.2 billion people
Lack of micronutrients	Malnutrition – vitamin and mineral deficiency	2 billion people
Unbalanced diet	Malnutrition – overweight and obese	1 billion people
Hunger	Lack of calories and protein	900 million people
Underweight children	Inadequate food intake and disease	126 million people
Vulnerability to disease	Weakened immune system	60% of deaths due to infections and parasites

many people. The challenge, therefore, is how to feed the ever-increasing numbers of people, especially when so many are born into real poverty.

One of the impacts of rural poverty is that young people, especially young men, leave their villages and go in search of employment in the towns. In Africa, as in many other parts of the world, nearly half the population now lives in urban areas, but work is hard to find and, as a consequence, crime is a big problem. Those left in the rural areas therefore tend to be the old, the sick, the young women and the very young. So the workforce for hard agricultural labour is limited.

It seems that amid all the excitement of technological advances, we have lost sight of the bigger picture. Rampant poverty, hunger and malnutrition in the non-industrial countries of the tropics are exacerbated by deforestation, land degradation and the misuse of water resources. One consequence of this is that we have now reached the point where the only sources of new land for agriculture are the small areas of remaining forest, which are critical for their important ecological and environmental functions. These include the mitigation of climate change. So, for many reasons we cannot afford to lose these forests as they play a critical role in the health of our planet.

Agriculture is said to be the source of 15% of the greenhouse gases (GHGs) responsible for climate change. Al Gore[3] has brought this to our attention, calling it an 'inconvenient truth'. Agriculture's energy cost has probably been underestimated, however, as the use of fossil fuels in agricultural production is often overlooked. Bill Mollinson,[4] the advocate of permaculture, has nicely demonstrated this issue with his example of the energy costs of producing an egg. He points out that assessments of the cost of egg production should include the costs of the energy used in growing the crops and catching the fish used in the manufacture of pelleted chicken feed – in other words, the mining of coal and iron, oil refining, power generation, transport of raw materials, iron smelting, manufacture of farm machinery, manufacture of chemicals, plastics, fertilizers, pesticides and veterinary medicines, cultivation of feed crops, fisheries, fish processing, feedstock pelleting, transport of manufactured goods, etc. These costs are all part of the equation for sustainable egg production, especially if – as must be the case – the environmental costs are taken into account in trying to resolve the food crisis. When we go on to consider the issues of global inequity, land shortages and poverty, it seems clear that we have multiple 'inconvenient truths'. We are guilty of complacency if we think that all is well in agriculture!

So, how do we dig ourselves out of the mess we are in? The political and trade issues are very important, complex and outside my area of expertise, so I hope others will address them. Leaving them aside, it seems to me that we could, if we put our minds to it, make rapid progress towards reducing hunger, malnutrition and poverty in the rural communities of tropical countries. This will, however, require us to rethink how we practise agriculture and particularly how we stimulate growth in the rural economies of the least developed countries. In particular, it seems evident to me that agriculture needs to shift from environmental culprit to environmental, social and economic saviour. To think about this sensibly, we first have to look at the problem in more detail.

The obvious stocks of natural capital that are critical for agriculture are fertile soil and the availability and purity of fresh water. We have already seen that productive land has become a scarce resource, but so too is water. Already 2.8 billion people live with water scarcity, 1.2 billion of these living in areas where water withdrawals exceed 75% of river flows. Together soil and water are critical for productive land and for healthy farmers. In addition, however, there is also the less obvious capital stock of biodiversity. We tend to think of biodiversity in terms of cuddly animals and beautiful birds, insects and flowers. But actually it is the complex interactions between all the organisms that make up the totality of wildlife – the microbes, the creepy-crawly bugs and the hungry predators, as well as the plants and the trees. They determine the health of ecosystems. Healthy ecosystems are dependent on the proper functioning of the nutrient, hydrological and carbon cycles, as well as the complex food chains and reproductive cycles of all the organisms making up an ecosystem.

Living unsustainably depletes these stocks of natural capital and upsets the delicate balance of nature, leading to the breakdown of ecosystem function, loss of soil fertility, erosion and pollution – problems that are especially serious in the tropics and subtropics. All of this is translated into the loss of productivity from crops and livestock (Fig. 2.1). This loss of productivity then leads to the serious social and economic issues, such as hunger, malnutrition and poverty, which in turn result in urban migration and even social unrest. All of these problems can be rolled together to constitute a gigantic threat to the social, economic and environmental sustainability of our world and our society and, perhaps, even our survival. How we practice agriculture is at the heart of the overall problem. The simplest way of thinking about this problem is that land degradation leads to poverty, and that poverty leads to land degradation, creating a downward spiral from which it is hard to escape, especially if you are a poor farmer in the tropics and subtropics, unable to invest in the inputs needed to increase productivity.

Generally, it is the desire to provide food for domestic use and to create income that motivates farmers to practise agriculture. In hard times these motives may lead to over-exploitation of the natural capital, but the decision to do this may also be influenced by external factors, such as: (i) the weather; (ii) fire; (iii) social disputes; (iv) politics; (v) population pressure; (vi) poor health; (vii) climate-induced crop failure; or (viii) the actions of outsiders. These factors may lead to decisions to clear more forest, overstock pastures, or to grow crops in ways that deplete soil fertility and increase the risk of pests and disease (Fig. 2.1).

Inevitably all these unsustainable practices lead to the need to improve soil fertility. Manure and composts are one source of nutrients, but the supply of these is often inadequate. If, as we have seen, the purchase of artificial fertilizers is not an option then ecosystem degradation and soil erosion are highly likely. This in turn leads to the loss of biological diversity above and/or below ground. This loss of the organisms

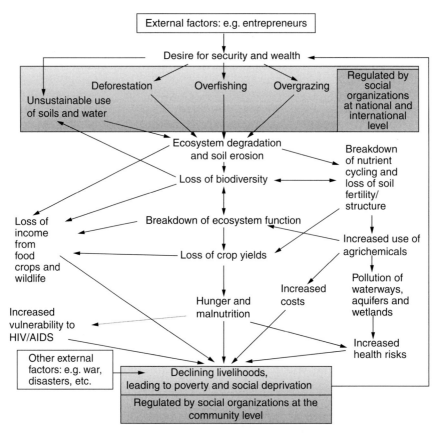

Fig. 2.1. Diagrammatic representation of the land degradation and social deprivation cycle illustrating the impacts of unsustainable use of natural resources on agroecosystems, farm production and, ultimately, poverty, malnutrition and hunger. This cycle is most evident in the tropics and subtropics. (Source: Leakey, 2010.)

responsible for the functioning of the agroecosystem eventually results in the loss of ecological resilience, with detrimental consequences for crop yield and income. In arid areas in particular, this can lead to failure to recycle nutrients and their transport to depths below the rooting zone, with their eventual passage down to the water table where they pollute groundwater. Declining livelihoods – often exacerbated by external factors, such as national and international policy, commodity prices, markets and trading opportunities – typically result in the cycle being repeated again. With each turn of the cycle the problems are compounded and the processes accelerated, reaching a point where all the environmental, social and economic problems overlap in a complex of land degradation and social deprivation (Fig. 2.1). I think the recognition that the key problem is that poverty leads to land degradation, and land degradation leads to poverty, makes it very clear that any solution we seek for better and more sustainable agricultural production has to simultaneously address both poverty and land degradation. All too often the solutions we have pursued in the past, it seems to me, have focused on the symptom – low crop yields – rather than these causes.

It should be clear from this description of this all-too-common phenomenon that efforts to reverse these processes will not be successful if policies and research initiatives do not try to simultaneously address the excessive consumption of natural resources that drives the downward spiral. Sadly, the consumptive philosophy of modern living extends from agriculture to economic development nationally and internationally, where the fundamental principle of recent decades has been reliance on the use and drawdown of natural capital, rather than reliance on the 'interest' derived from that capital for production. This aspect of globalization has impacted negatively on developing countries, and especially on their rural population, most of whom are farming households. What makes all the problems of poverty, hunger and malnutrition in developing countries so much worse is that the flipside in the rich industrial nations is over-eating, excessive materialism, greed, decadence and self-gratification.

International calls for renewed efforts to alleviate poverty and develop more sustainable land use systems have been ringing in our ears for about 20 years, even before the Rio de Janeiro (1992) and Johannesburg (2002) World Summits on Sustainable Development, and the World Food Summits in Rome (1996, 2006). In addition there have also been calls for 'a Doubly Green Revolution' by Professor Gordon Conway[5] and for 'a Green Revolution in Africa' by the Gates Foundation linked with the Rockefeller Foundation.[6] In addition, the United Nations (UN) has programmes in support of its Millennium Development Goal.[7] All these initiatives are extremely laudable, but the problem is they all fall short on how to actually practise sustainable agriculture. There is, however, general consensus internationally from international reviews like the *Millennium Ecosystem Assessment* (2009), *Global Environmental Outlook 4* (2007) and *The Comprehensive Assessment of Water Management in Agriculture* (2007) that, when it comes to environmental degradation, agriculture is a major culprit. These reports were followed in 2009 by the *International Assessment of Agricultural Science and Technology for Development*, which made a detailed study of the sustainability of agriculture and promoted the concept of multifunctional agriculture for enhanced environmental, social and economic sustainability. This report stated that 'business as usual' is no longer an option for agriculture – a statement that has also been reiterated by The Royal Society of London (2010).[8]

Adding all this dialogue together, there is overwhelming evidence that a change is needed. The question is still: 'How do we find a way to increase the productivity of farming systems that also creates a pathway out of poverty, malnutrition and hunger without increasing environmental degradation – or better still while reversing environmental rehabilitation?' This book is about finding a way of achieving this, especially in the tropics – and especially in Africa, where so much of the grief, anguish and agony are found.

We need a comprehensive solution. Our future, everybody's future, depends on it. It is dangerous to think that we are in control of our world and that food shortages are just blips in the normal pattern and fabric of life. Jared Diamond has reminded us in his historical analysis of the rise and fall of civilizations in *Collapse*[9] that others before us have been complacent and paid the price for over-exploiting natural resources. They consequently suffered the collapse of their civilizations. In the past some powerful civilizations were geographically small and could disappear without threat to our species. Now, however, as in effect we have become a single global civilization, this aspect of history will not repeat itself. We stand or fall together. We need to be very aware that, in the past, society's responses to environmental problems were highly significant in determining their impact on past civilizations.

These days society is consumption oriented. Nowadays we all expect to be white collar workers, driving big and expensive cars that we can leave parked outside our houses in up-market suburbs. It's a short-term philosophy peddled by politicians wanting an economic upturn before the next election and sold to us by companies managed by CEOs on big bonuses making profits for their shareholders. It's great. It means we can eat packaged and oven-ready meals while we watch one of hundreds of TV programmes punctuated by adverts encouraging us to consume even more. We don't have time to think about where this food came from, how it was produced or whether the farmer received a fair price for his labour. Maybe it is time for a rethink. The current global financial situation suggests that we have taken this lifestyle too far and we are not living sustainably.

It is perhaps useful to reflect on the fact that those civilizations that have already bitten the dust suffered the very same problems as the ones that are becoming increasingly common today. In particular, Jared Diamond identifies the following eight environmental problems as the common causes of a crash: (i) deforestation and habitat destruction; (ii) loss of soil fertility; (iii) erosion and salinization; (iv) mismanagement of water; (v) overhunting; (vi) overfishing; (vii) the replacement of native species with introduced species; and (viii) the impacts of overpopulation. Do these environmental issues sound familiar to us today?

Credible answers to the questions about the sustainability of agriculture that meet the food and livelihood needs of the wide range of people across the world are not easy to find. I believe it is a matter of finding: (i) how to mix and match; (ii) how to integrate modern scientific advances in agriculture and ecology with social sciences and culture; and (iii) how to recognize the different needs of people living in different circumstances. It is not a matter of 'one model fits all'. I believe we can do a lot better if we stand back and look at what we want from agriculture.

Notes

[1] Wangari Maathai (2010) *The Challenge for Africa*. Arrow Books, London, 319 pp.
[2] Julian Cribb (2010) *The Coming Famine: the Global Food Crisis and What We Can Do To Avoid It*. University of California Press, Los Angeles, California, 248 pp.
[3] Al Gore (2006) *An Inconvenient Truth: the Planetary Emergency of Global Warming and What We Can Do About It*. Rodale, Emmaus, Pennsylvania, 325 pp.
[4] William Mollinson (1988) *Permaculture: a Design Manual*. Tagari Publications, Tyalgum, New South Wales, Australia.
[5] Gordon Conway (1997) *The Doubly Green Revolution: Food for All in the 21st Century*. Penguin Books, London, 335 pp.
[6] Alliance for a Green Revolution in Africa (AGRA) with offices in Nairobi, Kenya and Accra, Ghana.
[7] Millennium Development Goals – eight goals for 2015 (www.un.org/millenniumgoals/).
[8] The Royal Society (2009) *Reaping the Benefits: Science and the Sustainable Intensification of Global Agriculture*. The Royal Society, London.
[9] Jared Diamond (2005) *Collapse: How Societies Choose to Fail or Survive*. Penguin Group (Australia), 576 pp.

Further Reading

Cassman, K.G., Wood, S., Choo, P.S., Cooper, D., Devendra, C., *et al.* (2005) Cultivated systems. In: Hassan, R., Scholes, R. and Ash, N. (eds) *Ecosystems and Human Well-Being*. Vol 1. *Current State and Trends*. Findings of the Condition and Trends Working Group of the Millennium Ecosystem Assessment. Island Press, Washington, DC, pp. 745–794.

Kiers, E.T., Leakey, R.R.B., Izac, A.-M., Heinemann, J.A., Rosenthal, E., Nathan, D. and Jiggins, J. (2008) Agriculture at a crossroads. *Science* 320, 320–321.

Leakey, R.R.B. (2010) Agroforestry: a delivery mechanism for multi-functional agriculture. In: Kellimore, L.R. (ed.) *Handbook on Agroforestry: Management Practices and Environmental Impact*. Environmental Science, Engineering and Technology Series. Nova Science Publishers, Hauppauge, New York, pp. 461–471.

Leakey, R.R.B., Kranjac-Berisavljevic, G., Caron, P., Craufurd, P., Martin, A., *et al.* (2008) Impacts of AKST on development and sustainability goals. In: McIntyre, B.D., Herren, H., Wakhungu, J. and Watson, R. (eds) *International Assessment of Agricultural Science and Technology for Development: Global Report*. Island Press, Washington, DC, pp. 145–253.

Journeys of Discovery in Agroforestry   3

> Tropical small-scale farmers eventually taught us that the integration of trees in agricultural landscapes has enormous untapped potential to benefit people and the environment. What remained was to deploy the power of science to accelerate the knowledge generation and productivity increases that would effectively exploit this potential.
>
> Dennis Garrity (2006) *World Agroforestry into the Future*.
> World Agroforestry Centre, Nairobi, Kenya.

> At this point agroforestry is a field of study which involves the combined application of ecology, economics, anthropology, agronomy, forestry, soil science, animal science, tree genetics, biometrics and other applied sciences. Out of this cauldron a recognized science may emerge, and what an exciting one it may be, given its broad interdisciplinary nature.
>
> Pedro Sanchez (1995) Science in agroforestry. *Agroforestry Systems* 30, 5–55.

During our descent into Nairobi, I looked out at the rift valley, recognizing Mount Longonot, which was close to my old school at Gilgil. We then circled south of the Ngong Hills losing altitude as we flew across the Nairobi National Park. A few giraffe were just visible out of my window, before we crossed over the Mombasa road and landed at Jomo Kenyatta Airport. In a way it was a home-coming, although more than 30 years had passed since it was my real home.

As I settled in to my new office at the International Centre for Research in Agroforestry[1] (ICRAF), I was aware that my understanding of what agroforestry was all about was fairly rudimentary. In essence I knew it was the inclusion of trees in farming systems, providing environmental services and a range of useful forest products. The potential seemed enormous, but from what I knew the reality was much more limited. I was aware that for about 15 years there had been research in many parts of the world, much of it by ICRAF, to develop what was called hedgerow intercropping or alley farming with nitrogen-fixing leguminous trees and shrubs to provide a simple way of restoring soil fertility. These trees, like peas and beans, produce root nodules that are colonized by nitrogen-fixing bacteria. These bacteria take up nitrogen from the air and store it, while some is also released to the soil. The stored nitrogen can be made available to crops by cutting the shoots and using them as mulch or 'green manure'. I was also aware that while these technologies worked well in experimental situations, they were not well adopted by farmers. I was not very excited about these things, but I had not said that at my interview! I thought that agroforestry should be much more than this.

My immediate task, therefore, was to learn as much as I could about agroforestry, as fast as possible. The best way to do that was to visit my agroforestry research teams in some 20 countries of Africa, South-east Asia and Latin America.[2] These teams were a combination of ICRAF staff[3] and local scientists from national research organizations. Seeing their work for myself would also give me the opportunity to expand my knowledge of tropical agriculture by talking to farmers about their needs and problems as, typically, our research was done in farmers' fields, rather than on research stations.

I soon discovered that one of the pleasures of interacting with these farmers was their cheerfulness and willingness to show me their farms. It seems that despite their enormous everyday hardships and their lack of material wealth, they have an inner peace and innate sense of hospitality that we have lost while we worry about our material possessions, our social positions, or other meaningless aspects of our hectic and pressurized modern lives.

In my forestry days I had looked at farms on newly cleared land and felt a sense of frustration. Now, I found that wearing an agroforestry hat I could accept that this land was farmland, and no longer forest. These farmers had to feed their families, but to my surprise I found they were often doing quite innovative things and that the future was not all gloom and despondency. They were practising agroforestry in its many forms – that is to say, using trees within their farms to provide fuel wood, food, medicines, timber and other essential products for everyday life. In other words, they were using trees to satisfy many of their basic needs so as to be reasonably self-sufficient. These trees were additionally providing environmental services such as shade, protection from wind, reduction of soil erosion and the maintenance of soil fertility. In some cases they were also actively growing forest trees for their products. This was encouraging. The other thing that was obvious was that there was room for further improvement using modern science; but before getting into that let's consider what agroforestry is all about.

There are many types of agriculture practised in Africa, where about 80% of the rural population are farmers. Much of Africa is dry and semi-arid, but there are also many other rainfall regimes from the wet lowlands with humid tropical rainforest to the cool, damp highlands with montane forest. Much of this variation is eco-regional, reflecting differences in total precipitation, altitude and latitude. In the areas of higher rainfall there can be a single wet season with a single dry season, or two wet seasons (one typically longer and wetter than the other) interspersed by two dry seasons (one usually cooler than the other). In the space that I have available, I will have to make sweeping generalizations. Across all these zones shifting cultivation was the traditional form of land use, but once again population pressures have led to significant changes. In many cases there are just no new areas to move to and so farming has become more sedentary and the land has become more degraded. Frequently, population pressure has also led to decreasing farm size (typically between 1 and 5 ha). Farm size declines as land is divided between the children of a family at the end of each generation. Sometimes this means that a single farmer has separated individual fields scattered across the area.

Land degradation is the common consequence of forest/woodland clearance and year-on-year cropping – in the worst-case scenario, without any period of fallow. In the absence of income to purchase fertilizers, declining soil fertility results in declining crop yields and a subsistence lifestyle – locked in poverty, malnutrition and hunger. While the option of growing cash crops to provide some income can be very beneficial in some areas, the financial returns from these commodities fluctuate at the whim of globally controlled markets. This tends to make the cultivation of these commodities an unreliable livelihood strategy.

When things get really hard, more or less the whole farm has to be planted with food crops such as maize. In the worst situations, the maize is so stunted that plants produce only one or two small cobs and overall yields can be less than 1 t/ha. In these situations, farmers do what they can to maintain soil fertility by adding either green manure or mulch gathered from natural vegetation from regularly pruned hedgerows. Alternatively, these shoots can be fed to livestock and their manure spread around the crops. Some benefits can also accrue from rotating maize with legumes like cowpea or beans, which are nitrogen-fixing plants. These activities may slow the decline in soil fertility, but seldom prevent it from happening eventually. When crops are stressed like this they are also more vulnerable to severe pests, like the parasitic witchweed (*Striga hermonthica*), as well as the cereal stem-borers (*Chilo partellus*). All these constraints are of course exacerbated by drought when the rains fail. Fortunately, there are also good years when things are not so bad, but the lives of these farmers are never easy. So, when this is the reality, how can agroforestry help these farmers?

For pragmatic reasons, I'm going to start this account of my voyages of discovery in southern Africa, as it was here that ICRAF's work developing better and more adoptable soil-fertility replenishment technologies was making exciting progress when I joined.[4] Hedgerow intercropping and alley farming had turned out to be too laborious for farmers, especially as the work load came at a time of year when the farmers were very busy with planting and other essential tasks. So, to get around this, the idea that was catching the attention of thousands of farmers in Zambia and Malawi was an 'improved fallow', which was based on nitrogen-fixing trees and shrubs such as *Sesbania sesban* and *Tephrosia vogelii*. The concept is simple. Small areas on the farm are taken out of production for 2–3 years and seed or seedlings of these leguminous trees established at about 10,000 plants/ha (i.e. 1 m × 1 m). They grow fast and produce a thick stand 2–3 m tall. Throughout this period they are doing their trick and enriching the soil with nitrogen. The trees are then cut down and allowed to decompose. They can also be dug into the soil.

In southern Africa I was shown striking examples of the impacts of this simple technique. These improved fallows had effectively restored nitrogen fertility in 2–3 years rather than in the 15–25 years required by traditional forest fallows and the maize yields went up from around 1 t/ha to about 4 or 5 t/ha (Fig. 3.1). This is a substantial gain, but unfortunately does not bring maize yield up to the 10 or so t/ha that can be attained with the use of chemical fertilizers. Nevertheless, this improved fallow technology greatly improves food security and so is critically important to millions of farming households. The problem is that most African farmers are unaware of this approach to improving their maize yields.

By the time of my first visit to southern Africa, the main focus of the ICRAF team was on dissemination of the technology through local farmer groups and training schools. In parallel with this there was work to help farmers produce and distribute seeds to keep pace with the demand from other farmers. Fortunately, most leguminous trees are prolific seed producers and the seeds are easy to collect and to germinate. Farmers can therefore help one another to get started and make a big step upwards in

Fig. 3.1. Maize without (left) and with (right) a *Sesbania sesban* fallow in southern Africa.

terms of their ability to meet their needs for calories from staple food crops. The main requirement for wider dissemination of this easy approach is therefore to provide help so that more farmers can adopt these simple soil-fertility enhancing technologies.

Of course one system of soil fertility enrichment does not necessarily work everywhere, as both the physical and social environments vary from place to place. Consequently, where constraints exist, adaptive research is required to find ways around specific constraints to adoption. For example, in the humid forest zone of West Africa land degradation has the additional problem of soil acidity, but trials with a range of local and exotic legumes found species, such as *Calliandra calothyrsus* (Fig. 3.2) that were effective as improved fallows under these conditions.

Social constraints have similarly been overcome. For example, when farmers cannot afford to take fields out of production for 2–3 years to grow an improved fallow, the technique has to be adjusted. One approach is to grow the crops and the nitrogen-fixing shrubs simultaneously. At first sight this seems unlikely to work, as you would imagine that there would be fierce competition between the crops and the shrubs. However, with the correct timing, the reality is that the maize will grow fast during the wet season while the shrubs remain small in their shade, but then as the maize cobs ripen and the crop plant dies back the shrubs can start growing. They then continue to grow and fix nitrogen in their root nodules through the dry season, so enriching the soil for the next cropping season. This has been called 'relay cropping' (Fig. 3.2). Like the improved fallow, the end result of 2–3 years of this is better crop yields from improved soil fertility. Another candidate species for relay cropping is *Gliricidia sepium*. Together, these different nitrogen-fixing trees and shrubs have been dubbed the 'fertilizer trees'.

Fig. 3.2. Improved fallow with *Calliandra calothyrsus* (left) and relay cropping with *Sesbania sesban* (right) in Cameroon.

Another fertilizer tree of interest in Africa and being developed by ICRAF at the moment is *Faidherbia albida*, which is often cherished by local people because of its special soil-improving qualities. It is unusual as it sheds its leaves in the wet season when crops are growing and produces new leaves in the dry season when there are no crops in the ground. This obviously means that both above- and below-ground competition with crops is minimized. Consequently, it can be grown as a big tree among staple food crops like maize, millet and sorghum. This is being described as 'evergreen agriculture'.

The natural vegetation of much of eastern, central and southern Africa is woodland savannah called the Miombo woodlands. These woodlands extend from Tanzania and the southern Republic of Congo in the north, to Zimbabwe in the south and across the continent from Angola, through Zambia to Malawi and Mozambique, covering more than 2.7 million km^2. Prior to land clearance, these woodlands constituted the most extensive tropical seasonal woodland and dry forest formation in Africa, and had one of the world's highest levels of botanical biodiversity. Unfortunately, current rates of deforestation in the Miombo woodlands are among the highest in the world, creating shortages of wood fuel as well as the soil fertility problems we have already discussed.

Interestingly, these Miombo woodlands contain about 50 tree species producing edible fruits (e.g. *Strychnos cocculoides*, *Azanza garckeana* and *Vangueria infausta*) and ten or more species producing high quality, slow-growing hardwoods (e.g. *Khaya nyasica*, *Pterocarpus angolensis*), while others are suitable as sources of fodder. Various other foods, such as mushrooms, bulbs, roots, young shoots, caterpillars and insects, are also obtained from Miombo woodlands. Not only are these nutritious, but they are marketable. One of the indigenous fruit trees in this area is the marula (*Sclerocarya birrea*), which in addition to providing traditionally important fruits and nuts is used to brew a culturally important local beverage. At the time of my first visit to southern Africa there was interest in seeing if better use could be made of this natural resource of indigenous fruits, which were generally harvested from wild trees scattered at low density in farmland.

At this time there was also emerging evidence from studies by Rhodes University, Grahamstown, South Africa[5] and ICRAF that showed that farmers in southern Africa consumed quite large quantities of these fruits and that they are also a source of income for local people. Fruits from marula trees were sold at roadside stalls and could raise household income above US$822/month in the fruiting season. This was a considerable sum in an area where smallholder farmers have very little, if any, income. This makes these trees very attractive to local farmers. Interestingly, we found that in some areas individual marula trees in the farmland had 'pet' names that link them with a family. This indicates the importance of these trees.

The importance of the indigenous fruits and their occurrence in farming systems certainly sparked my interest.

In East Africa, ICRAF research[6] near Kisumu in western Kenya showed another important aspect of improved fallows. The roots of the trees penetrate the soil deep below the crop rooting zone, where they act as a safety net for nutrients that are being lost from the cropping system into the groundwater: as for example happens after heavy rainfall. The uptake of these nutrients by the tree roots returns them to the aerial shoots, where they are used to make new leaves and branches. Then in the dry season when the leaves wither and fall, the nutrients are recycled back to the soil surface as leaf litter, where they are once again available to crops.

Together the research on improved fallows in southern Africa and in Kenya has also led to great advances in the understanding of the underlying chemical, physical and biological processes associated with soil fertility depletion and replenishment. These and other contributions to soil fertility management were recognized by the award of the World Food Prize to Professor Pedro Sanchez in 2002.

So far I have probably given the impression that nitrogen-fixing species are only grown for their ability to improve soil fertility. However these species can also be used for other purposes, so that nitrogen fixation is then a secondary benefit. First, they can be used to provide another environmental service – the capture and efficient use of limited rainfall and erosion control. When planted along the contours of hillsides, trees and/or hedges can greatly reduce runoff and erosion. In addition, through the accumulation of organic matter from leaf litter and the turnover of roots, trees also help to make the soil structure more open. This improves the infiltration of water into the soil profile, where it is available to crops.

In an example from Machakos,[7] hedges were found to double the amount of water infiltrating into the soil on a 14% slope, and to reduce runoff by 75%. This outcome reduced soil erosion from 19 to about 2 t of soil/ha depending on the tree species used. The 'cost' of this improved water capture was the loss of 27% of this water to the atmosphere as transpiration from the trees' leaves. Overall, however, there was a net benefit. Results like these indicate that, with appropriate knowledge, there is potential to develop cropping systems that make much better and more efficient use of available water resources. The question, however, is how to make it a practical and adoptable technology?

The benefits of trees in contour plantings are even greater on the steeper, terraced hillsides of Rwanda, Burundi and western Uganda,[8] where contour trees and hedgerows additionally play the important role of stabilizing the terrace banks. The treeless views in this upland area are breathtaking, but if you look beneath the green veneer, you see that the crops are failing and the soils are eroding where the poorly constructed terraces are washed away. The priorities here are first to reduce the erosion by stabilizing the terrace risers with trees and hedges and secondly to improve the soil fertility at the back of the terrace, where the subsoil has been exposed when the topsoil was scooped forward to form the front of the terrace.

For the farmers in these areas, one of the greatest constraints is access to tree seedlings. The species found to be best for both terrace stabilization and the restoration of the yields of beans and sorghum through enhanced soil fertility on the terrace, were the Central American trees *Alnus acuminata* and *C. calothyrsus*. Consequently, one of the activities of ICRAF in this area was teaching local people how to develop their own community tree nurseries. The scale of this need is enormous, as literally hundreds of miles of hedges are needed to stabilize the terraces of even a small community. In the longer term, of course, these trees will also provide much needed wood fuel and potentially many other products.

While contour hedges play this important role in erosion control, other research at Machakos, at the boundary of the semi-arid zone and the central highlands of Kenya (rainfall = 750 mm), was finding that other trees are also important in water harvesting. For example, when the timber tree *Grevillia robusta* was grown at 3 m × 4 m spacing among maize, the trees intercepted 25–45% of the rain, while the maize captured 30–35%. This reduces the erosive power of the rain and again improves infiltration to the soil. While trees and crops compete for water in the top 50 cm of the soil profile, the reduced runoff means that more water is available. At greater depths, where trees actually obtain about 60% of their water requirements, the tree roots can capture the water escaping to the groundwater. So, in both of these ways trees act like a sponge and mitigate the undesirable impacts of heavy rainfall.

If for a moment we think about the incidence of serious cases of flooding and landslides internationally, we can see that there is a huge need for protecting vulnerable hillsides with contour trees and hedges. In 2010 alone, there were the serious floods in Pakistan and India, as well as in Australia. Extraordinarily there is seldom any mention in the press that denuding hillsides is the cause of these disasters. Nor is there any recognition that the loss of property and life caused by this flooding could be mitigated by sensible use of erosion control measures such as contour hedges and tree planting.

As we have just seen, trees compete with crops for water, and indeed for nutrients and light. This competition can and often does result in lower crop yields. At first sight, this seems to be a problem, one that might make you think that agroforestry was a bad idea. However, as we have seen there are numerous environmental benefits and so ultimately these all have to be considered to see if the net effect is positive or negative. Interestingly, this kind of environmental accounting usually results in a positive overall impact. There is, however, another factor, one that definitely tips the balance in favour of the trees. It involves more straightforward economic accounting. As I have mentioned many times, trees can produce useful products. In this instance, if the tree products are more valuable or useful than the lost crop yield then the overall economic returns exceed the financial losses.

The potential range of products from trees is almost as great as the number of useful tree species, and covers timber, wood for fuel and making numerous day-to-day tools, fruits, nuts, fodder and medicines. This is what I really wanted to find out more about as I travelled around visiting my research teams. Progressively over this and the next few chapters we are going to see how this aspect of agroforestry can be developed. Ultimately, however, it is when the environmental benefits and the economic and social benefits are drawn together that we really see what agroforestry is all about.

So, we have seen that nitrogen-fixing plants enrich the soil and, like other trees, also provide environmental services. Now we will look at the production of some tree products in more detail, starting with fodder, again from nitrogen-fixing leguminous trees. In East Africa, especially in Kenya, a dairy industry is emerging on the slopes of Mount Kenya, around the Embu District. We are not talking about large dairy herds and sophisticated milking parlours. Typically the farmers here are smallholders with only 1–2 ha of land growing a range of food and some cash crops, like coffee or tea. They have one or two milking cows, and perhaps a goat or a sheep, which are all kept in sheds and stall-fed with fodder brought in from the fields – a cut-and-carry system. Commonly, this fodder comes from hedged nitrogen-fixing trees like *C. calothyrsus* or tall grasses like napier grass (*Pennisetum purpureum*) planted on earth banks on field boundaries, or on contour ridges to reduce erosion. However, just as in dairy farms in Europe and other parts of the world, grass alone is not sufficient to attain maximum milk yields. Here in Kenya, however, the expensive cattle concentrate pellets that are used in Europe to boost milk yields are not an affordable option. ICRAF's work here[9] with scientists from the Kenya Agriculture Research Institute and Kenya Forestry Research Institute was to see how to use more nutritious tree fodders to improve dairy production.

The outcomes of this research are both exciting and important for this fledgling industry. It was found, for example, that a 400 m *Calliandra* hedge could produce enough fodder to feed two cows (90 kg/day) in the dry season (mid-December to mid-March). This raised the milk yield by over 300 l. In other words, 3 kg of homegrown *Calliandra* tree fodder was equivalent to 1 kg of purchased dairy 'concentrate'. Today, the programme has greatly expanded and the fodder trees are being widely adopted. In much the same way, studies in Burundi have found that tree fodder was also beneficial for the enhanced production of goat meat through the dry season.

The East African highlands have high population density – often over 500 people/km². In my youth these areas had been severely deforested to make way both for large-scale plantations owned by expatriates, as well as for the needs of local farmers needing to feed their families. Today, however, the Kenyan landscape contains many more trees. This seems to be contrary to the generally accepted rule that high population pressures are associated with treeless landscapes. So, what has happened here? The answer is that farmers have planted a wide range of tree species, some indigenous, like *Cordia africana* and *Croton megalocarpus,* and some exotics from other parts of the world, such as the Australian trees *G. robusta* (silky oak) and *Eucalyptus saligna* and *Eucalyptus grandis* (eucalypts or gum trees), Central American *Cupressus lusitanica* and *Cupressus macrocarpa* (Mexican and Monterey cypress) and Asian *Mangifera indica* (mango). These trees are frequently grown on farm boundaries, as contour plantings, or scattered across the farm. In addition to producing wood or other products, boundary plantings have the added role of preventing encroachment by neighbours and conferring security of tenure to the land.

Something interesting is going on here! From the above, it is clear that an agroforestry culture has emerged in the Kenyan Highlands. This has changed the landscape from large-scale monocultural plantations of tea and coffee to highly diversified small farms. Interestingly, land covered by many small farms has a greater abundance and diversity of trees than areas with larger farms,[10] indicating that households that have to support a family on a small plot of land spread their risks. Typically these small farms are occupied by the older and youngest members of the family, while the young men and some women are working (or seeking work) in urban areas to earn 'off-farm' income to support themselves and the family. The tree-based farming is not labour intensive and so is more easily managed by the elderly.

This tree-planting phenomenon is interesting as it indicates that when the pressures on land are already very high, population growth actually stimulates a reversal of deforestation through a more sustainable approach to farming – agroforestry. Okay, the vegetation cover is not as diverse and complex as natural forest and not all the species are indigenous, but the agroecosystem is very much more diversified, and the landscape is a mosaic like we see in natural landscapes. Interestingly, this tree-planting revolution is also found in a few other high population density areas of Africa, such as Burundi. In times when climate change is becoming a serious problem, and agricultural technologies are not promoting sustainable land use, this is very good news – the concept of 'more people: more trees' is something not recognized by most policy makers!

I noticed another unexpected activity when I revisited the higher elevations of Nyeri and Embu. The farmers were planting high-value indigenous timber trees like *Vitex keniensis* (Meru oak). When I questioned farmers about why they had planted slow-growing high-quality timbers like Meru oak I was told that they saw them as a long-term investment to pay for big events such as marriages, funerals, etc., as well as for the education of their children and grandchildren. When farmers start to think like this, there is real hope for the continuation of tree planting in the landscape! Seeing the planting of Meru oak was especially pleasing to me, as my father[11] had actively promoted its growth during his career as a forest officer in Kenya. Furthermore, a few years earlier I had co-supervised a Kenyan PhD student, Paul Konuche, in Edinburgh. He had worked on the light requirements for the regeneration of Meru oak before going on to become the Director of the Kenyan Forestry Research Institute.

We've seen that trees in farmland can provide fodder and timber, but trees are also grown for fruits, nuts, medicines, building materials, and wood for carving. By producing tree products for their own consumption, farmers have the security of reduced risk of depending on purchased products, as well as having an alternative source of income in difficult times.

The medicinal value of tree products is something that is not appreciated by many people in the industrialized countries of the world, yet if you look in almost any book on the taxonomy of trees from any country of the world, you will see a long list of the medicinal uses of bark, roots, leaves, fruits and seeds. This is especially true for tropical countries where the local people are still regular users of these 'herbal' medicines, as a walk through most local markets will quickly illustrate.

In the East African highlands, as in Cameroon, one of the important medicinal products comes from the bark of a montane tree *Prunus africana* (pygeum) trees (Fig. 3.3). The bark of this tree is the source of the drugs used to treat benign prostatic hyperplasia – the swelling of the prostate gland which sends men over 50 running to the toilet to urinate. Unsustainable harvesting of the bark kills the tree and with a huge demand for the bark by pharmaceutical companies this can lead to very serious loss of forest. An additional concern is that its fruits are also an important source of food for wildlife, so the over-exploitation of this tree species for its bark is also a matter of interest to conservation organizations because some endemic birds and monkeys are becoming rare in montane Africa. Cultivating this tree therefore has potential to be good for poor farmers, good for wildlife and good for the pharmaceutical industry – an interesting and surprising triumvirate!

Throughout Africa, the scarcity of fuelwood in areas without trees is often as ubiquitous as the problem of soil infertility. Often the lack of fuelwood for cooking leads to farmers having to burn dung and maize stalks. This use of dung as fuel means that it is not available to be used as a source of nutrients in manure. Thus, all alternative sources of fuel are beneficial. One option is to use the larger stems available at the end of a 2–3 year improved fallow. These stems are a valuable source of low-grade

Fig. 3.3. Bark harvested from pygeum (*Prunus africana*) for the medicinal plant market in Cameroon.

fuelwood (10–35 t/ha). However, on the other side of the same coin, leguminous trees, such as the Australian *Acacia* trees, can be deliberately grown for fuelwood but have the secondary benefit of improving soil fertility. In Tanzania, at Shinyanga for example, short-rotation village woodlots were found to produce 9–13 t of wood/ha over 3 years. Combine this with soil fertility enrichment and you have a very useful improvement to land use in this dry and barren area of Tanzania.

Moving on to the drylands and to the Sahel in particular, we see some new challenges for agriculture. Here, in marked contrast with the small dairy-cattle units in East Africa, there are large nomadic herds of goats and cattle that range around the waterholes and grazing areas quite independent of the sedentary farmers – well, for part of the year at least. The lives of nomadic pastoralists in Africa are regulated by a very complex set of tenurial rights: rights of access to, and use of, land and water resources, and even to individual tree species. These are traditional and customary rights implemented by community-based regulations. These rights vary from place to place, and between ethnic groups, and are often not adequately recognized by the legal systems of individual nations. This makes generalizations very difficult. However, in the Sahel the sedentary farmers typically have sole access to their land in the wet season, while their crops are growing. But after harvest, in the dry season, the pastoralists have the right to graze their animals in the farmland. In the past, when population densities were not so high, this was sustainable, but now the farming system is threatened as the periods of natural fallows are too short to allow the scattered trees to regenerate naturally and the herds make it difficult for the sedentary farmers to grow trees for fodder or other products. Lengthy and sensitive discussions are needed between the two communities, often mediated by local chiefs, local authorities and government, if any changes or restrictions are to be agreed without conflict. In some cases special committees are established for this purpose.

The agricultural landscape of the Sahel is dominated by what is called the 'parklands' – scattered large trees (40–80/ha) producing fruits, nuts and dry-season fodder (Fig. 3.4). Traditionally, these well-recognized and useful trees were retained in the farmer's fields when the woodland savannah was cleared during shifting cultivation to plant crops such as millet, sorghum and groundnuts. Typically, the crops grow well in close proximity to the trees. This is perhaps a combination of some protection from the hot sun, and nutrients from leaf litter, bird droppings, etc. Poverty, water scarcity and low soil fertility are the biggest problems of the very large population of people living in the parklands. Livestock contribute about 70% of farm income. However, the availability of dry-season fodder is a major constraint to livestock production.

Because fodder is in short supply much of the year, it is not unusual to see men on bicycles loaded with large quantities of tree shoots peddling towards town destined for local markets. This laborious activity is

Fig. 3.4. Sahelian parklands dominated by *Vitellaria paradoxa* (shea nut).

an indicator of the need for a less damaging alternative. To overcome the shortage of dry-season fodder I was shown research to develop fodder banks with both indigenous species, typically leguminous trees such as *Pterocarpus erinaceus*, and exotics like *G. sepium*.[12] These experimental fodder banks were managed as hedges or regularly cut back stumps, often within the protection of thorny hedges. These can be very productive, with a hectare producing about 4.5 t/year valued at US$630. This income is 2.5 times the average annual per capita income of peri-urban farmers. These leguminous fodder trees also fix nitrogen, so they have the secondary benefit of soil fertility restoration.

Traditionally fields of crops were protected by fences made from the stems of the previous year's sorghum or millet crop. By the dry season, these fences were decomposing and so were no longer an obstacle to the nomads' herds. Both the environmental and the product-producing roles of the parkland trees are threatened by the demise of the parklands, so other research that I was shown was examining how to replace the traditional fences with thorny hedges so that the crops can be protected and a few tree seedlings could be grown at widely spaced intervals to restore the parklands. There were several different candidate species for hedges, one of which, *Zizyphus mauritiana*, also produces edible fruits. However, this solution was not just a matter of producing tree seedlings, planting and managing the hedges to enclose the farmers' fields, as it was clear that enclosing fields would set the sedentary farmers and the nomads in conflict over access to grazing and to waterholes. If hedges are going to be used to enclose the fields of sedentary farmers then corridors must be left for herds of livestock to move around the area and the lost grazing must be provided in other ways. This will require both landscape planning and a negotiated new relationship between the herdsmen and the farmers. So, one of my first lessons was that it is always very important to look at the social implications of all ideas about how to develop solutions to agricultural problems. Maybe fodder banks could also provide an element of the solution – a bargaining chip – in discussions about how to meet the needs of pastoralists if hedges are to be introduced extensively.

The trees forming the Sahelian parklands of West Africa are typically indigenous fruit and nut trees, such as shea nut or karité (*V. paradoxa*), néré (*Parkia biglobosa*) and baobab (*Adansonia digitata*), which are retained in the field systems of the sedentary farmers. The products from these trees are used by local people as a source of food on a daily basis, and are traded in local, regional and international markets. Many of them are highly nutritious, being rich in minerals, vitamins, essential amino acids, etc. Shea nut, for example, is one of the most common trees of the Sahelian parklands (Fig. 3.4). One hundred kilograms of its fruit give about 5 kg of shea butter, with an oil content of 46–52% (33% unsaturated and 67% saturated). Shea butter is used as baking fat and margarine and is increasingly used in edible products, such as patisserie and confectionery and as a cocoa butter substitute in chocolate. Shea nut oil and butter are also exported for use in cosmetic and pharmaceutical industries.

The seeds of néré, on the other hand, are fermented to make a strong-smelling food condiment called 'soumbala', which is eaten in sauces with sorghum/millet dumplings throughout the year. It is rich in protein (40%), lipids (35%), linoleic acid and vitamin B_2. The sale of néré fruits can increase household income by US$270/year. This may not sound much, but for people earning less than US$1/day is not insignificant. Similarly, the yellow, floury pulp around the seeds of baobab pods is also rich in vitamin C (ten times more than oranges). This sugary pulp can be eaten raw or mixed with water to produce a refreshing drink, or fermented into an alcoholic beverage. The seed kernels contain 12–15% edible oils and more protein than groundnuts, and are rich in lysine, thiamine, calcium and iron. In addition, baobab leaves, which are eaten as spinach, are rich in vitamin A. To obtain these leaves, these huge trees have to be climbed. To get around this, one of the ideas being tested by ICRAF projects was the possibility that baobabs could be hedged for leaf production, in much the same way as tea plants are managed. On one of my visits to this region I saw a newly established baobab hedge. It looked pretty promising, but it is interesting to speculate how it will do and what it might look like in 50–100 years. Will the leaves be harvested from hedges with vast stems like those of mature free-standing baobabs, or will the regular management of the hedges keep the stems small?

Because I was familiar with the lowland humid tropics of West and Central Africa from my earlier activities in tropical forestry, I concentrated my early travels to places with which I was less familiar. However, in contrast with the dry areas, the vegetation of this region is green and lush and there is relatively little involvement in livestock production. Here, the main food crops include cassava, yam, maize, plantain, cocoyam and groundnuts, and fruits like pineapple, mango, avocado, citrus, guava and papaya. In addition, well-recognized cash crops like cocoa, coffee and oil palm are often integrated within these smallholder farms. With this range of food crops, one might expect that this region should have fewer problems than the drier areas of Africa. However, despite the lush growth and the greater diversity of food crops, the common problems of soil degradation and poverty are just as serious here as elsewhere in Africa. For example, the soils of the humid forest zone of West Africa are infertile, deficient in nitrogen and often prone to acidification. Soil acidity has two undesirable effects on fertility. First, it leads to phosphorus being more strongly attached to soil particles, making phosphates less available to crops; and secondly, it results in aluminium toxicity. Seeking species that could be used as improved fallows therefore had to test their tolerance of acidity and capacity to fix nitrogen and neutralize these soils.

Oil palm and rubber are also grown as large-scale plantations owned either by government or multinational companies. Cocoa farming, on the other hand, is predominantly a smallholder enterprise in which cocoa trees are grown under the shade of either secondary forest or planted shade trees. Cocoa is an understorey tree in its natural Amazonian habitat. Many of these smallholder farms are now relatively old and have an upper canopy of indigenous trees for timber as well as fruits and nuts, forming a forest-like farming system or agroforest. In addition to its diverse canopy, it also has a well-diversified ground flora. These cocoa agroforests contain a wide range of exotic and indigenous species (Fig. 3.5) in

Fig. 3.5. Cocoa agroforest in Cameroon with maize in the foreground.

approximately a 50:50 ratio. This is also true in south-east Nigeria, where the population is very high (>1000 people/km^2). Here, 29% of the cultivated area is in diverse and complex species mixtures, which produce 59% of the crop output. In monetary terms, the outputs of these agroforests, many of which are tree products, are five to ten times greater than those of crop fields.

For me, as we saw from my visit to Kumba market in Chapter 1, it was the realization of the importance of indigenous fruits and nuts in the farms and in the markets that had the greatest significance in terms of how the role of agroforestry could be further developed. The importance and value of these local fruits and nuts such as *Dacryodes edulis* (safou) and *Irvingia gabonensis* (bush mango) in local and regional markets is substantial. The fruits of the former are roasted as a vegetable, while the kernels from the nuts of the latter are an important source of a polysaccharide, which forms a much appreciated glutinaceous food-thickening agent for local soups and stews. We will look at these and other traditional food species in more detail in Chapters 5 and 6.

From my previous trips to Asia for forestry research I was aware that there is a very different feel about the countries and I was aware that the Green Revolution had been more successful here than in Africa. Furthermore, I had heard of, but never seen, the complex agroforests of South-east Asia, and especially of Indonesia.

So it was with a sense of anticipation that I travelled to Indonesia to see the work of my colleagues in the Sumatran agroforests. I went straight to Krui in the south-west corner of Sumatra, not too far from the island of Krakatoa, which is famous for appearing out of the sea during a volcanic eruption. On the way we stopped only briefly to see some soil fertility improvement experiments near Tanjungkarang.

Krui is an area with gently rolling hills covered in what looked like dipterocarp forest (Fig. 3.6), with paddy rice in the valley bottoms (Fig. 3.7). The innovation introduced by these farmers was to plant the hillsides cleared for agriculture with damar (*Shorea javanica*) seedlings at the same time as they planted their food crops (typically dry land rice, peppers, coffee, bananas, etc.). So, in fact what I was seeing was a dipterocarp agroforest, and not a natural forest. The damar trees were planted at approximately 50 trees/ha, together with other forest species (cinnamon, rubber and fruit/nut trees like duku (*Lancium domesticum*), durian (*Durio zibethinus*), candlenut (*Aleurites moluccana*), etc.). Damar is a large dipterocarp timber tree that also produces a resin that the farmers tap, harvest and market to the paint and varnish industry. As the agroforest grows and matures, wild species colonize and fill the gaps between the trees.

In this way, the farmers get a dry land rice crop on the valley sides for the first 2–3 years while the trees are establishing. The rice crop is then followed by the peppers, root crops, bananas and coffee until about year

Fig. 3.6. Agricultural landscape in Indonesia – complex multistrata agroforest adjacent to crop fields.

ten, when the tree crops close the canopy. Tapping the damar trees for resin begins when the trees are about 10–15 years old and this is more or less a monthly process that generates income throughout the year. The damar and other tree crops will be productive for at least another 40 years. When their productivity declines they can be felled for timber, and the cycle repeated. Throughout this cropping cycle the farmers are also able to gather fruits, vegetables, herbs, spices, medicinal products from the wild species in the agroforest, so enhancing food and nutritional security. Likewise the gathering of construction wood, firewood, thatching and basketry materials helps to meet many of the day-to-day needs of the households. Typically these agroforests provide between 50 and 80% of the villages' total agricultural income. In Indonesia, agroforests produce 70–80% of the commercially

Fig. 3.7. Paddy rice with cinnamon and rubber agroforest in Sumatra, Indonesia.

traded damar resin and roughly 95% of the marketed fruits and nuts. In addition, they collectively provide 63% of the cinnamon, nutmeg, candlenut and a significant proportion of the trade in rattans and bamboos.

The overall result of this innovation is that the farmers have replaced the unproductive forest fallow with an agroforest that is, in effect, a commercial fallow, a capital asset full of commercially important tree species that will be productive for many years. So by enriching the fallow, these farmers have developed a more-or-less permanent forest resource, which is at least as productive as a natural forest as it is contains whatever mix of species the farmer planted to meet the local market opportunities. The combination of the paddy rice in the irrigated valley bottoms, the food crops from young agroforests and tree products from the mature agroforests meets most of the day-to-day needs of the communities as well as generating income from trade. As a result the livelihoods of these farmers are well above those still practising shifting cultivation or just growing paddy rice. In 1994, about half of the population was involved in damar production and related activities like harvesting, transport and trade. Thus these agroforests build a substantial resource in the forest products and create new market opportunities which contribute to off-farm employment and community livelihoods in many ways. So, by growing these agroforests these farmers have improved their own lives and developed a sustainable farming system that avoids many of the negative impacts of modern agriculture.

This visit to Indonesia opened my eyes to some very different and interesting aspects of agroforestry.[13] When visiting Krui, I saw that local farmers have used their intuition and local knowledge to respond to the loss of forest and forest products in a very novel but entirely logical way. This allowed the farmers to change from shifting cultivation to a sedentary lifestyle, without suffering the downward spiral of land degradation problems resulting from a shortened period of fallow. In other words, they had avoided all the problems of population growth by focusing more on the trees and the forest products than on the staple foods of the Green Revolution. Yet in their rice paddies they could adopt and benefit from improved varieties and technologies arising from the Green Revolution.

In addition to the productivity of the agroforests on the valley sides they minimized risk from over-reliance on a few crops and, most impressively of all, they created a very sustainable farming system that was beneficial to both the environment and the livelihoods of the local farmers. On their own initiative these farmers had led a silent, unfunded revolution of their own – running in parallel with the Green Revolution. These farmers with little or no formal education, no political clout and no capacity to voice their ideas, had found a novel and highly effective solution to their need to sustain life amid the disappearing forest. In effect their solution was to replace the natural fallow of the shifting cultivator with a high-value fallow containing a substantial number of commercially important tree species that provide them with marketable products.[14] By 1994 when I visited Krui, over 80% of the villages scattered along 120 km of the Pesisir coast in south-west Sumatra had damar agroforests.

During my second trip to Sumatra I visited Padang, where I saw cinnamon and rubber agroforests developed in the same way as I have described for damar. Rubber, of course, used to be a large-scale plantation crop tapped for its latex, but here in Indonesia it was now grown by 1.3 million smallholder families farming 2.5 million ha as a component of mixed agroforests. By 1997, these rubber agroforests were the source of

73% of the nation's rubber (= 25% of world rubber), with a value of US$1.9 billion. Farmer income from rubber agroforests (US$2300 from 2 ha) has been reported to be slightly in excess of a monoclonal rubber plantation of the same size, and is thus one of the most profitable land use systems for smallholders in Sumatra.

Similar agroforestry approaches, often involving other tree species, are not confined to Sumatra. This is not the declining relic of some former civilization, but is the response of farmers to their increasingly difficult situation. In Indonesia the total area of these productive agroforests had increased to about 3 million ha by 1995. As the studies of these Indonesian agroforests became known, it was realized that similar systems exist elsewhere – mostly in Asia, but also in a few African and Latin American countries (Table 3.1). In hindsight, it is also possible to see the West African cocoa farms as a somewhat similar approach to the development of agroforests. This wider distribution illustrates that the concept has worldwide relevance. It also emphasizes that farmers in many parts of the world are knowledgeable about their indigenous trees, as well as about the use and trade opportunities for their products.

One of the many things that impressed me about these Indonesian agroforests was that they combined remarkable social and economic benefits with a stream of environmental benefits, like those of natural forests. Agroforests have been found to contain about 60–70% of the animal species – from small insects and invertebrates to birds and monkeys – that are found in a natural forest. This contrasts with about 3% in an oil palm plantation. In addition, farmers in Krui have indicated that the water supply for rice fields in the valley bottoms has been considerably improved in quantity and regularity since the agroforests on hillsides replaced shifting cultivation. In contrast to most conventional agricultural systems, the high biomass of these agroforests means that, like a natural forest, they sequester large quantities of carbon and so are very good land use options for the reduction of GHG emissions to mitigate climate change. Furthermore, the soils of agroforests have also been found to act as sinks for methane emissions from the paddy fields.

Earlier, I mentioned the development of tree-based farming systems in East Africa and the concept of 'more people: more trees'. Looking out of the aeroplane window while I was flying from Padang to Jakarta, I realized that I was seeing the same phenomenon. After take-off the rice paddies and agroforests of Padang become forest with evident logging tracks as we flew over the Bukit Barisan mountain ridge. Then the forest thinned and slash-and-burn agriculture became clearly visible, with smoke rising from small cleared areas in a mosaic of forest and crops. This then gave way to large areas of sugarcane in the totally treeless landscape of the coastal plain of Jambi and Selatan Provinces in east Sumatra. But then, as the population increased again, farm size decreased and trees came back

Table 3.1. Examples of complex agroforests from South-east Asia and elsewhere. (Source: Michon et al., 2005.)

Continent	Country	System
Asia	Indonesia	Benzoin resin gardens in North Sumatra
		Rattan gardens and shifting agriculture in East Kalimantan
		Rubber gardens in Sumatra
		Damar agroforests in Pesisir, Lampung, Sumatra
		Fruit and timber agroforests in Maninjau, West Sumatra
		Candlenut agroforests of South Sulawesi
		Fruit forests (lembo) of the Mahakam river: East Kalimantan
		Illipe-nut forests (tembawang) in West Kalimantan
		Durian forests of West Kalimantan
		Fruit and cinnamon agroforests of Kerinci
		Fruit agroforests of Jambi and Palembang
		Sugar palm and salak agroforests of Bali and Lombok
		Fruit gardens of West Java
		Bamboo gardens of West Java
		Spice (nutmeg and clove) and nut agroforests of the Moluccas: in association with banana groves, coconut and nut-producing *Inocarpus* trees, and *Canarium* trees
	Laos	Cardamom gardens of southern Laos
		Benzoin fallow gardens in northern Laos
		Coffee–fruit–forest gardens of southern Laos
	Thailand	Miang tea agroforests, with fruit trees and other species
Africa	Tanzania	The Chagga home gardens of Mount Kilimanjaro include coffee and banana groves under *Albizzia* species
	Ethopia	Arabica coffee forests with cultivated species and self-regenerated forest trees
	Nigeria and Cameroon	Cocoa agroforests with *Irvingia gabonensis, Dacryodes edulis, Coula edulis, Cola* spp.
Latin America	Mexico and Honduras	Maya home gardens
	Brazil	Fruit agroforests of Carnero island in the Amazon estuary
		Brazil nut groves and palm forests of the Amazon region

into the landscape, until on the approach to Jakarta I saw homegardens filled with a range of useful tree species. Looking at this, I pondered on the fact that the increasing diversity of the agroecosystems below me was almost certainly associated with higher living standards.

These observations of 'more people: more trees' are not exclusive to the tropics; similar patterns of land use change and demographics can be seen as you fly in to London across the south of England, possibly with the same changes in wealth. On this flight to Jakarta, I remembered too that my CEH colleagues Professor Fred Last and Bob Munro found that the number and diversity of woody plants in the city of Edinburgh were greater than those in the surrounding Scottish countryside.

Prior to joining ICRAF, my trips to Latin America had been limited to Costa Rica where I had visited and worked at Centro Agronómico Tropical de Investigación y Enseñanza (CATIE) on the development of vegetative propagation techniques for the timber tree *Cordia alliodora*. As a result of this experience, I was aware that CATIE staff had been active in agroforestry research including interesting systems in which the nitrogen-fixing tree *Erythrina poeppigiana* was grown as a regularly cut-back tree in coffee plantations under *C. alliodora*. I had seen these systems and appreciated that they were a nice example of a simple agroforestry approach to some extent mimicking the different strata of a tropical forest.

Now with staff in Mexico, Peru and Brazil,[15] I had a chance to broaden my experience of the New World. However, our teams in Latin America were relatively new and had been posted to sites where agroforestry certainly had the potential to play an important role in rural development, but at the time of my visits there was not too much to see on the ground. So, for example, I visited some ejidos in the Yucatan Peninsula where tree planting activities were just starting. These were government-sponsored agricultural cooperatives to resettle rural and urban landless families in village communes. I also visited some highly developed homegardens with a wide range of fruits, vegetables and tree crops. It was interesting to discover here how knowledgeable the farmers were about the ecological interactions between the various species.

In Peru, after a short flight from Lima over the Andes into Yurimaguas on one of the headwaters of the Amazon, I visited agroforestry experiments set up many years earlier by staff of North Carolina State University. One of these experiments compared the productivity and economic returns from: (i) shifting cultivation; (ii) intensive agriculture with high inputs of agrichemicals to maintain soil fertility and control pests; (iii) peach palm plantations; and (iv) complex agroforests based on fruit and timber species. Interestingly, the data indicated that the agroforest was economically by far the most attractive land use option. In effect this was because the agroforests were productive every year, for many years, with minimal expenditure on agrichemicals. In contrast, shifting cultivation was only productive for 2–3 years and then unproductive during the fallow phase, while the expenditure needed to maintain intensive cropping permanently, year on year, rose dramatically. The other option, the peach palm plantation, was unproductive for a number of years until the palms matured. It was certainly very enlightening see the data comparing the economics of these different approaches to agriculture so clearly enumerated.

My visits to Brazil were in marked contrast to all the others that I have mentioned. I visited areas being settled under the Trans-Amazonia Migration Programme in Rondonia and Acre. In both of these states the government had sponsored large-scale colonization schemes for migrants from other areas in Brazil. These migrants were therefore not familiar with agricultural practices for the rainforest areas or with the local species with economic potential. Under these schemes, migrant households were allocated 100 ha blocks of forest laid out in a grid, with two farms back to back, such that in a satellite image the countryside seems to be laid out in a fishbone design, a chequerboard of forest and pasture.

Under the terms of the scheme, the migrant farmer was not supposed to clear the forest from more than half the block. In Rondonia the area had already been colonized for some time and in many farms more than half the block had already been cleared of forest and it was dominated by pasture for beef cattle, much of it already severely degraded. Consequently, many of the original 100 ha settlements had merged as more successful farmers, typically large corporate cattle-ranching companies, had bought out those who had gone broke. In Acre, in contrast, the settlement scheme was more recent and its 100 ha landholdings were largely intact and still practising shifting agriculture.

ICRAF's fledgling agroforestry projects in these areas were aimed at: (i) the rehabilitation of already degraded pasture land by establishing silvo-pastural systems (growth of trees within pastures) to both diversify and increase pasture land productivity and large-scale improved fallows to reclaim degraded and abandoned pastures; and (ii) improving the level of land use management on the forest margin to prevent the slide towards degradation, by the creation of productive forest-like systems involving both annual and perennial crops, such as indigenous trees producing fruits and nuts. Amazonia is very rich in indigenous species with edible fruits, most notably peach palm (*Bactris gasipaes*), many of them quite popular and traded both fresh and processed as ice-cream flavours, etc. Peach palm fruits and 'heart of palm' are canned and bottled for export to the USA – a US$100 million industry.

When I thought about these visits to Amazonia in relation to adoption of agroforestry and the 'more people: more trees' scenario that I mentioned earlier in East Africa and in Indonesia, this really brought home to me the importance of farmers knowing the local tree species, as well as the importance of small farm size for sustainability. If you have to feed your family on 2–3 ha, you manage it much more carefully and sustainably than if you have 100 ha. I find, however, that people get quite upset when I say the problem with Amazonia is that there are not enough people in it!

Brazil is so vast that on one trip we hired a small plane to show some dignitaries around the area and to get a good view of the huge area of forest and the expansion of agriculture. Interestingly, these small planes use widened lengths of road as airstrips – the main indicator of this use being a windsock on the roadside. So, you circle around while the pilot checks the road for a gap in the traffic before landing. On one of these stops we flew in and were picked up to see some agroforestry experiments set up by Dr Erick Fernandes of North Carolina State University. They were a series of complex species mixtures set out in different configurations around mahogany trees, which were growing very well. Because of my former interest in mahogany, which is so vulnerable to pest attack in monocultures, this was very interesting for me. Interestingly, the trees in these species mixtures seemed to be relatively free from shoot-borer attacks.

On a second trip to Peru, 2 years later, I visited Pulcallpa and we went out on the River Ucayali, one of the upper tributaries of the Amazon, to visit some villages. It was still early morning and there was a mist hanging over the river as the sun rose in the sky. The river was in spate and carrying quite a heavy load of debris from erosion further upstream in the foothills of the Andes. We were stopping every now and then, as we went up the river, to ask farmers about the tree species they appreciated and to get information about the flowering and fruiting behaviour. Just as we were leaving one site a police launch hailed us, telling us to stop and we were asked what we were doing. It was the 'Drugs Squad' asking routine questions in their efforts to halt the flow of the drugs trade. They were satisfied by the explanations of our skipper and with a burst on the throttle of their powerful motors they roared off into the mist leaving a considerable wake behind them. We didn't give them any more thought until we stopped at a village for lunch and found people talking about an accident. It turned out that not long after leaving us the police boat had hit a submerged log and been overturned. Unfortunately, it seems that several of the policemen had not been able to swim and they had been drowned in the swollen river.

Now, many years later, I have visited other areas of Brazil, Bolivia and Costa Rica and seen some excellent examples of complex agroforestry systems in practice. This illustrates the potential to further develop agroforestry solutions to land use problems in Latin America. In addition, some of the cocoa agroforests are much more developed than the West African cocoa farms; more like the Indonesian agroforests. This also illustrates the potential to create better cocoa agroforests in Africa.

Having seen the work being done by my ICRAF colleagues, I was now really hooked on the idea of agroforestry. Up until that time, I had not really understood what it was all about. Seeing and hearing about it first-hand convinced me that it was indeed an important land use concept with great relevance to agriculture, especially in the tropics. However, much of what I had seen was about intercropping, the mixture of trees and crops in rows across a field and things like boundary plantings, wind breaks, living fences and hedges, fodder banks, scattered trees in cropland or pastures, shade trees in plantation crops like cocoa and coffee, improved fallows, small woodlots, orchards and homegardens. In all these situations agroforestry was a useful and important agronomic approach to the integration of trees in farming systems that improved the food security of rural households and went some way to improving their livelihoods. So, yes, agroforestry was indeed about producing both products and environmental services for millions of people. However, my travels had also led me to believe that agroforestry could be much more than that.

Notes

[1] The term 'agroforestry' was coined in the mid-1970s when the International Development Research Centre (IDRC) of Canada created the International Council for Research in Agroforestry (ICRAF).

[2] We had staff based in 14 counties of Africa (Kenya, Uganda, Rwanda, Burundi, Tanzania, Zambia, Malawi, Zimbabwe, Cameroon, Nigeria, Burkina Faso, Niger, Mali and Senegal), three in Asia (Indonesia, Thailand and the Philippines) and three in Latin America (Mexico, Peru and Brazil).

[3] Totalling nearly 300.

[4] Led by Dr Freddie Kwesiga and Professors Jumane Maghembe and Pedro Sanchez.

[5] Sheona Shackleton *et al.* (2002) *Southern African Forestry Journal* 194, 27–41.

[6] Led by Drs Roland Buresh, Bashir Jama and Paul Smithson.

[7] Led by Professor Chin Ong, Drs Meka Rao, Ahmed Khan and Nick Jackson.

[8] Led by Drs Don Peden, Amadou Niang and Antoine Kalinganire.

[9] Led by Drs Mick O'Neill, Rob Patterson, Steve Franzel and Ekow Akyeampong.

[10] Kindt, R., Simons, A.J. and van Damme, P. (2004) Do farm characteristics explain differences in tree species diversity among western Kenyan farms? *Agroforestry Systems* 63, 63–74.

[11] Professor Tony Young has described him as one of the early advocates of agroforestry in his book *Agroforestry for Soil Conservation* (1989, CAB International, Wallingford, UK), on account of a paper he presented at the African Soils Conference in the Belgian Congo in 1948.

[12] Led by Drs Edouard Bonkoungou, Mamadou Djimdé, Elias Ayuk and later Drs Amadou Niang and Zac Tchoundjeu.

[13] Led by Drs Dennis Garrity, Meine van Noordwijk, Genevieve Michon and Hubert de Foresta.

[14] Michon, G. *et al.* (2005) *Domesticating Forests: How Farmers Manage Forest Resources*. Institut de Recherche pour le Développement, Paris, France; Center for International Forestry Research, Bogor, Indonesia/The World Agroforestry Centre, Nairobi, Kenya, 203 pp.

[15] Led by Drs Marcelino Avila in Mexico, Dale Bandy and Julio Alegre in Peru and Carlos Castillo in Brazil. Later John Weber transferred to Peru and Jeremy Haggar joined in Mexico.

Further Reading

Cooper, P.J.M., Leakey, R.R.B., Rao, M.R. and Reynolds, L. (1996) Agroforestry and the mitigation of land degradation in the humid and sub-humid tropics of Africa. *Experimental Agriculture* 32, 235–290.

Leakey, R.R.B. (1998) Agroforestry in the humid lowlands of West Africa: some reflections on future directions for research. *Agroforestry Systems* 40, 253–262.

Michon, G., *et al.* (2005) *Domesticating Forests: How Farmers Manage Forest Resources*. Institut de Recherche pour le Développement, Paris/Center for International Forestry Research, Bogor, Indonesia/The World Agroforestry Centre, Nairobi, Kenya, 203 pp.

Ong, C. and Leakey, R.R.B. (1999) Why tree–crop interactions in agroforestry appear at odds with tree–grass interactions in tropical savannahs. *Agroforestry Systems* 45, 109–129.

Sanchez, P.A. and Leakey, R.R.B. (1997) Land use transformation in Africa: three imperatives for balancing food security with natural resource conservation. *European Journal of Agronomy* 7, 15–23.

Sanchez, P.A., Buresh, R.J. and Leakey, R.R.B. (1997) Trees, soils and food security. *Philosophical Transactions of the Royal Society Series B* 352, 949–961.

Diversity and Function in Farming Systems

4

> Despite its success, our system of global food production is in process of under-cutting the very foundation upon which it has been built. The techniques, innovations, practices and policies that have allowed increases in productivity have also undermined the basis for their productivity.
> Stephen R. Gliessman (1998) *Agroecology: Ecological Processes in Sustainable Agriculture*. Ann Arbor Press, Chelsea, Michigan.

> The essence of successful diversification is to maintain profitability by growing high-value products with an adequate market demand; while maintaining the landscape as both a mosaic of different farming systems at different stages in their cropping cycles and, where possible, to develop integrated crop mixtures.
> Network for Sustainable and Diversified Agriculture (NSDA) (2004) *Vision Statement*. NSDA, Cairns, Queensland, Australia.

It has been estimated that approximately 1.2 billion farmers practise agroforestry, while about 1.5 billion people (over 20% of the world's population) use agroforestry products. From my travels seeing a wide range of different agroforestry systems, I realized that agroforestry is more than just an agronomic practice that restores soil fertility and produces tree products in farmers' fields. It is also applied ecology or, more accurately, applied agroecology – the ecology of farming systems. This means, therefore, that it could be expected to also deliver ecological functions over and above such environmental services as erosion control, water infiltration, provision of shade, etc. that we saw in the last chapter. Environmental services are basically physical processes, while ecological functions have to do with the biological processes that make ecosystems dynamic and that regulate the balance between different organisms. This ecological balancing act is all about regulating the interactions between organisms throughout their life cycles and along their food chains. So, this process is an altogether higher order of magnitude in the way life self-regulates and creates a balance between species. It is this balance that confers ecological sustainability in different types of vegetation, landscapes or land uses.

The reason that I was excited about this realization is that modern intensive agriculture is notoriously destructive of all these processes. First of all because it typically reduces the diversity of species and cuts the dominant plant species to one – a monocultural crop. Secondly, it uses agrichemicals to replace some of the key agroecological functions by the use of pesticides to prevent pathogens, pests and weeds from taking over control of this dominant invader. In other words, the agrichemicals try to stop the natural food chains in their tracks, so that the crop is unaffected by other organisms. This means that conventional high-input agriculture is always fighting nature, and in the tropics this fight can be a fierce one as there are so many organisms struggling to impose some natural 'law and order'. This would not be too serious if it were happening on a small scale, but agriculture occupies nearly 40% of the land surface – so it has a huge 'footprint' on the global environment.

Some agriculturalists are very critical of the idea of introducing ecology into agriculture. The argument seems to be that ecology is not 'good' science. However, a more serious examination of the complexity of ecological interactions reveals that agroecology is really the next big scientific frontier and a massive challenge to modern science. A really good understanding of agroecology, and in particular the role that trees and diversity play in the promotion of agroecological function, could certainly revolutionize how we produce our food. Unfortunately, we are a long way from this level of understanding at the moment. In the meantime agroforestry seems to be a good way to deliver some ecological and environmental sustainability in agricultural landscapes.

At the time of my arrival in ICRAF, the working definition of agroforestry was 'a collective name for land use systems and practices in which woody perennials are deliberately integrated with crops and/or animals on the same land-management unit'.

What I had seen as I visited my staff, especially in Indonesia, was that agroforestry is actually applied agroecology and you can think of the diversification of the farming system with different tree species as the inclusion of *planned* biodiversity. However, from an ecological viewpoint, this planned diversification then promotes the inclusion of *unplanned*, or associated, biodiversity. The latter is composed of all those organisms that move in to fill the above- and below-ground niches among the planted trees and crops. It was clear to me that the addition of trees to farmland starts a process of increasing biological diversity and that as new niches are created for wildlife in the farmers' fields, so the agroecosystem enters an ecological succession. This succession would be similar to that found in natural vegetation recovering from some severe damage – ranging from a tree fall, to a landslide or even a volcanic eruption. In other words, it is a process of repair to environmental damage and as it progresses it restores the ecological balancing act. This view of agroforestry allows the benefits of tree planting described in the last chapter to be seen as entry points into this succession, which ultimately creates mature agroforests like those I saw in Indonesia – which in my experience are among the most environmentally sustainable farming systems in the world. Hopefully you remember that they were also pretty good socially and economically, as well.

With the above in mind, we see that when farmers manage the agroecosystem they can decide whether or not the farm should remain in the pioneer stage by repeatedly cultivating and replanting the field with a new annual crop or, alternatively, to allow certain parts of the farm to progress towards a mature agroforest. A decision to halt the succession at an intermediate point is also possible. The latter two options of course create a landscape mosaic of increasing biological diversity and complexity leading, importantly, to environmental benefits. With increasing scale, the agroecosystem becomes more stable and starts to meet many of the needs of the global community for what in development jargon are called 'international public goods and services': things like watershed protection, reduced emission of greenhouse gases (GHGs), and the retention of biological diversity. As we have seen, these can be achieved without that ever being the farmer's specific intention; he or she is seeking reliable food security and income generation for their household.

With these ideas I wrote an article entitled 'Definition of agroforestry revisited'. By suggesting a new definition, I was trying to move agroforestry from its predominant focus on agronomic practices to a much wider vision in which agroforestry practices place trees in the landscape to maintain or restore ecological functions and services, as well as to produce marketable goods. I therefore defined agroforestry as 'a dynamic, ecologically based, natural resources management system that, through the integration of trees in farms and in the landscape, diversifies and sustains production for increased social, economic and environmental benefits for land users at all levels'. ICRAF adopted this definition in 1997.

With the new definition we can look in a bit more detail at some of the ecological functions, and particularly at the role of biological diversity and succession. In intensive monocultures, the unplanned biodiversity is kept to a minimum by the use of expensive tillage equipment and agrichemicals, as this diversity is seen as undesirable and a threat to production. Rather than fighting nature to be productive, agroforests work with nature to be productive. This involves the creation of mixed farming systems based on trees producing marketable products – the planned biodiversity. The trick for profitable agroforestry production therefore is to plant as many commercial species of herbs, vines, shrubs and trees in as complex an agroecosystem as possible.[1] Ecological theory suggests that the more planned diversity there is in the agroecosystem, the greater the quantity of unplanned biodiversity that will colonize the system, and that this diversity will continue to rise as this system progresses through an ecological succession. In a few instances, some components of this unplanned biodiversity will also be marketable and can add to the overall economic value of farm products.

If we consider a tropical rainforest that has a very high biomass, it can only get sufficient nutrients for growth and survival by very rapidly recycling the nutrients held in its biomass. So as leaves, twigs, branches and tree trunks fall to the ground, they are rapidly invaded by the unplanned biodiversity – the numerous worms, termites, bugs, beasties and micro-organisms that gnaw, chew and digest the biomass, absorb the nutrients, defecate, die and rot down, so that the nutrients are made soluble and can be drawn back up into the forest plants for their continued growth. While this is going on at the forest floor, there are also insects, birds and mammals up in the forest canopy that are also eating the leaves and fruits, as well as each other, and again defecating and dying, and so making nutrients available again even more rapidly. In addition there is also a network of roots and fungal filaments below ground to trap and recycle the nutrients back into the vegetation, so preventing them from being washed out of the soil by heavy rain. Some of these fungal filaments have special relationships with the roots of the plants they colonize. They are known as mycorrhizas and the relationship is symbiotic – in other words, beneficial to both the plant and the fungus. The fungal filaments help the plants to scavenge for nutrients and water. In exchange the fungi can benefit from the sugars coming down from the leaves to feed the roots. These processes are the driving forces of the nutrient and carbon cycles – the foundations of soil fertility and the reduction of carbon dioxide emissions to the atmosphere.

Mycorrhizal fungi are very important for the tree establishment, survival and growth. They are also very vulnerable to environmental disturbance. For example, when a forest is cleared there is an almost instant crash of the populations of forest fungi and they are rapidly replaced by fungi associated with the pioneer plants and weeds. It can then take many years under a forest plantation before these pioneer fungi are once again fully replaced by populations of the forest fungi. The absence of the appropriate fungi makes it more difficult to establish forest trees on cleared sites. The appropriate mycorrhizal populations can be extremely important for tree establishment in degraded arid land sites. To overcome these problems tree seedlings can be deliberately inoculated with the appropriate fungus in the tree nursery.

All of the above is much more important in the tropics than in the temperate zone. This is because tropical ecosystems are much more complex.

In addition, the soils in cool temperate climates are more fertile. This is because in cool climates the organic matter breaks down more slowly and so accumulates in the soil. In contrast, the soils of tropical environments are geologically old and low in mineral fertility. This is exacerbated by the combination of high temperatures, moisture and the high biodiversity of tropical soil organisms, which together lead to rapid breakdown of organic matter so the soils are shallow. Actually most of the nutrient stock in tropical ecosystems is in the plants – the biomass – and not in the soils. These differences between the tropical and temperate zones make agroforestry more important in the tropics.

There are several other aspects of agroecology that we need to consider. First, scale is important, as within natural ecosystems there is a hierarchy of organisms living at different scales. So, a bacterium in the soil may never move more than a few centimetres. It may be eaten by a nematode that travels a few metres, which will itself be eaten by a small mammal running around on the forest floor covering several kilometres. Although these food chains function reasonably well at scales as small as a hectare, the most efficient function only occurs when the top predators, such as an eagle or a jaguar can play their part. This requires a population of individual top predators each with a territory of many square kilometres if they are to breed satisfactorily.

Most plants, the bigger ones at least, are of course anchored to the spot by their roots. However, their populations can travel as seeds, often in the intestines of birds and animals, or in rivers. In addition, plant genes are carried around the landscape as pollen on the wind, or on insects, birds and mammals. Both seeds and pollen transport can be relatively local or long distance. So, plant species vary in the area required to support a viable population. The importance of this is that, if we are to find out the true impacts of agroecological factors on productivity and profitability, it is critical that the work is done at the appropriate scale.

In an agricultural landscape, the achievement of sufficient scale for top predators can probably be provided by a landscape mosaic that includes food crops, tree crops and natural vegetation; especially if there are some corridors of perennial vegetation providing connectivity between the mature components of the agroecosystem. In practical terms, landscape mosaics provide diversity in time and space – due to the location, configuration and duration of different species in the landscape. Part of this variability results from farmers applying different farming systems and management practices in accordance with their personal preferences. These will be influenced by: (i) differences in farm size; (ii) the wealth of the farmer; (iii) access to market; (iv) the tenure systems; (v) the availability and price of labour; and (vi) the availability of other sources of income. In ecological terms, this additional source of variability is desirable.

In the context of climate change, we perhaps need to recognize the impact of agriculturally induced land degradation and ecosystem dysfunction. When land is cleared for agriculture and cultivated, two of the repositories of stored carbon are adversely affected: (i) the woody vegetation, which sequesters carbon dioxide as carbohydrates and cellulose in woody perennial tissues; and (ii) the organic matter in the soil. The decomposition of vegetation and soil organic matter as a result of aeration and the burning of cut vegetation releases many different GHGs – most notably carbon dioxide – altogether contributing about 15% of global atmospheric emissions attributed to agriculture. Much of this could be prevented by the large-scale integration of trees into farming systems. Estimates by the World Agroforestry Centre suggest that carbon could be increased from about 2 t/ha in severely degraded land up to 90–150 t/ha in a dense agroforest over an area of about 900 million ha worldwide.[2]

Concerns about climate change are not just due to global warming. Increased incidence of drought is a real concern of many farmers. Again forest clearance is implicated. Professor E. Salati and Dr P.B. Vose measured the way water is recycled around the Amazon Basin[3] and determined what proportion of the moisture arriving at one site was passed on to the next site. Let me try to explain it simply. When clouds come in off the ocean they carry moisture that subsequently falls as rain. In the first rainfall event some of the rain reaches the ground, but some is evaporated off the leaf surfaces and carried back up into the atmosphere. Of the rain reaching the ground, some runs off into streams and finds its way back out to sea, while the remainder percolates into the soil. Some of this soil water is then taken up by plants to be converted to sugars during photosynthesis or is transpired out of the leaf through the stomata. This too will get back into the atmosphere. The moisture that was returned to the atmosphere enters phase two and is blown inland to fall in the next rainfall event, where all the same series of processes occur. This cycle can be repeated many times across a continent. When deforestation occurs the proportion reaching the ground is increased and much less is recycled back to the atmosphere, so reducing the quantity of rain that can fall in the next rainfall event, with serious implications for rain-fed agriculture inland. The question, of course, is: can the hydrological processes in agroforests sufficiently mimic natural forest to have beneficial impacts on rainfall downwind in continental interiors? We don't know the answer.

The problem at the moment is that we do not have enough hard scientific data to provide adequate knowledge of all the ecological, hydrological and environmental processes at play to be able to convince the sceptics of the value of this ecological approach to agriculture. This research has, however, been started and many of the complex relationships are becoming better understood. Nevertheless we need the science community to do much more to unravel the complexities of how agroecosystems function.

I have been concerned about the lack of good scientific knowledge about the potential ecological benefits of agroforestry for many years. I hope it is clear from what I have said that there is good reason to believe that integrating trees within farming systems should minimize the environmental damage that agriculture does when we grow food in ways that do not maintain the agroecosystem functions. So this brings me back to the need to restore and rehabilitate degraded land and then to make it productive by growing tree crops for their products.

How much diversity do we need to restore agroecological function? If we are honest, we do not know because we do not really understand the role of diversity, even at the small-plot scale, for any crop anywhere in the world. In the face of this reality, I have tried to initiate some experimental work, using cocoa production as a model system.

My first opportunity to think about this arose when I was asked to formulate a study in Brazil. This came about because of a serious crash in cocoa production due to a disease called witches' broom, caused by the fungus *Crinipellis perniciosa*. Traditionally, cocoa (*Theobroma cacao*), an understorey species of Amazonia, has been grown at low density under the shade of secondary forest. This system is called 'cabruca' in Brazil, a form of agroforestry. However, back in the 1960s it was decided to try to increase productivity of cocoa by growing it without the shade of canopy trees. Part of the reason for this was the idea that the removal of the shade would reduce the incidence of fungal diseases by lowering the humidity of the cocoa trees' environment.

Over the first 10–15 years the removal of the shade did increase Brazilian cocoa production very rapidly, so that by 1983 it was over 400,000 t/year from about 6060 km^2. This led to a change in government policy to encourage farmers to reduce the overhead shade. Unfortunately there was inadequate recognition that the diversity of trees in a secondary forest provides many more ecological services and functions than just shade. The crash came when witches' broom struck. This was paralleled by a fall in the price of cocoa from US$3632/t to US$700–800/t. This double disaster led to farmers either abandoning their farms, or replacing their cocoa with alternative enterprises. This experience with cocoa in Brazil clearly demonstrates the dangers of overlooking the importance of maintaining agroecosystem functions and the over-dependence on a single crop.

When I was asked my opinion about this crash I suggested that one approach to overcoming the problem was to replace the role of the unproductive shade trees with indigenous trees that produced marketable products. I suggested that this should be done experimentally in order to obtain some understanding of the numbers of species in the planned biodiversity, and their configurations in the farming system that are necessary to restore the agroecosystem function and to make the system profitable again. Such an experiment could also produce information about the unplanned biodiversity and hence generate fundamental knowledge about the nature of sustainable production. Consequently, I proposed an experiment aimed at answering the following three questions:

1. Can an ecologically acceptable density of cocoa be made economically acceptable to farmers by diversification with other cash crops?
2. What combinations of cocoa, shade trees and other trees/shrubs can create a functioning agroecosystem that is also profitable for farmers?
3. How many trees/shrubs (the planned biodiversity) are required to create sufficient ecological niches above and below ground (the unplanned biodiversity) to ensure that cocoa grows well, remains healthy and produces beans on a sustainable basis?

From the answers to these questions we could arrive at some idea about the ecologically acceptable number of cocoa plants per hectare; and what density of cocoa and other components of the system are ecologically and economically sustainable.

My idea was to create a series of cocoa agroecosystems with increasingly higher cocoa densities under a constant density of shade trees of either two or 24 different species. This, I thought, would allow us to study the effects on cocoa production of the relationships between the planned and unplanned biological diversity when grown under the different physical, chemical and biological environments created by the different densities of cocoa and the different configurations of shade trees. I expected that the environmental parameters which would change most dramatically under these conditions would be: (i) the humidity and movement of air through the canopy; (ii) the amount of light reaching the ground flora; (iii) the recycling of water, nutrients and carbon; (iv) the incidence of disease and pest organisms such as fungi and insects; and (v) the incidence of beneficial insects, birds and mammals. All of these things would also affect the overall production and economic returns from cocoa as well as all the other marketable products from the planted shade-tree species. As you can see this experiment is a bit ambitious. The outcome of this idea was what is called a Nelder fan, developed with the assistance of a colleague, Dr Ron Smith.

By planting trees down the 'spokes' of the fan, the distance between the rows of plants increases as the fan spreads out. So, at the centre of the fan the cocoa plants are at the high density of 1120 trees/ha. Then as the fan spreads out the distances between the cocoa planted along the radiating rows is gradually increased so that the density of the planting declines, until at the outer edge it is only 100 cocoa trees/ha. On its own, this part of the experiment allows us to examine the effect of changing the cocoa density. We would certainly expect that somewhere along this gradient of

100–1120 trees/ha we would find a density that would give the maximum yield as a result of optimizing the availability of light, water and nutrients. This increase in the spacing between plants could be expected to also result in changes in the penetration of sunlight and its impact on temperature, as well as changes in the humidity of the air due to better air circulation. These changes in cocoa density should also lead to more physical space for colonization by other wild plants between the cocoa, and consequently to the increase in the number of other organisms present and dependent on the increasing diversity of plant species. Together these effects on the physical environment might also result in changes in the incidence of pest and disease organisms (such as the witches' broom fungus).

The second part of this design is then superimposed on top of the cocoa fans. It is a layer of trees providing the shade – or more accurately the two layers of shade provided by canopy trees (40/ha) and sub-canopy trees (80/ha). Here the density of these trees is constant across the whole experiment. This is achieved by reducing the distance between the concentric rings of trees providing the shade. In this way it is hoped that the amount of shade cast by these trees is constant. Finally, the experimental design then has one additional feature involving the number of different tree species forming both the canopy and the sub-canopy shade. Basically we have two treatments. In one there is a single canopy tree species with a single sub-canopy tree species, and in the other there are eight canopy species and 16 sub-canopy species.

My expectation was that by having two very different levels of species diversity in the planted shade trees (planned biodiversity) would demonstrably affect the availability of ecological niches for a very wide range of colonizing unplanned biodiversity moving in to fill the niches in these different habitats. The principle being tested by this is that species diversity in the planted trees should lead to increased interaction between species forming the unplanned biodiversity in terms of the levels of predation, herbivory, pathogen/host relations, etc. – all the things that maintain the balance between species in natural ecosystems. For this to happen quickly, I hoped that this experiment would be planted on land recently cleared from forest, or at least close enough to forest for wildlife to invade the experimental site.

The plan was for a multidisciplinary team of plant physiologists, ecologists, agronomists, chemists, micrometeorologists and economists to collect data to address the following areas of research:

- the temporal changes in productivity and economic returns from cocoa, as well as the various other timber and non-timber forest products;
- the development of a microclimate, and its suitability for the spread of pests and disease (relative humidity, evaporation rates from surfaces, duration of surface wetness) and for other flora and fauna;

- the development of unplanned biological diversity above and below ground (plants, small mammals, birds, insects, soil arthropods, mycorrhizal fungi, etc.) as new niches are created in agroecosystems with differing planned biological diversity;
- the changes in soil fertility, carbon sequestration, etc.; and
- the effects of the microclimate and biotic environment on the quality attributes of cocoa beans from diversified agroecosystems.

My hope was that this experiment would start a process leading to a better understanding of sustainable land use worldwide – a process based on the principles of species and economic diversification and their impacts on the ecological functioning of agroecosystems. It's a grand aim. Sadly, however, for reasons beyond my control, the experiment has never been established.

A few years later a second opportunity arose to do something similar in Vietnam. The idea this time was to do what is called a replacement series. In its simplest and most common form, a replacement series involves planting two species in various mixtures – for example 100:0, 75:25, 50:50, 25:75 and 0:100. In this experiment, my idea was that cocoa would be one of the two species, but the second species would in fact be replaced by a mixture of four cash crops. Then, as before, canopy trees would be overlaid over the design to provide the shade. In this case the canopy shade was to be provided by five canopy species and five sub-canopy species, each producing marketable products. The upper storey trees and sub-canopy trees/shrubs are at the same spacing (9 m × 9 m) and each of them is surrounded by either eight plants of cocoa, mixtures of cocoa and four cash crops, or just the four cash crops without cocoa (3 m × 3 m). The modular structure of the canopy ensures that the physical properties of the multistrata agroforest (i.e. the three storeys of the canopy) remain constant across the experiment. In addition to the replacement series, each of the different species was also grown as a monoculture. This was to allow the comparison of production and income per unit area from a monoculture with that of the integrated species mixture.

Specifically, this second experiment seeks to examine the relationships in mixed-species cocoa agroforests between spacing, microclimate, the planned and unplanned biological diversity, the incidence of pests and diseases in cocoa, and the overall production and economic returns from cocoa and the other companion trees. The study adds an assessment of the impacts of different configurations of companion species grown in differing proportions within the canopy. Again, a multidisciplinary team of researchers would be needed to seek answers to the same sort of research questions. Once again the experiment was never established. Hopefully, one day donors and research organizations will make the necessary commitments to implement large, long-term, multidisciplinary experiments like these.

So far we have been looking at the role of diversity in ecology, but for smallholders production is also about meeting all their daily requirements for food and a range of other products – so it is about production and income. In a subsistence situation, seeking 'economies of scale' by buying up your neighbours' farms is not an option. So what can you do to support your family in the absence of social services, savings, investments or even employment?

As we saw in Chapter 3, it is generally true that diversification results in the lower yield of the primary product. However, if five planted species are producing marketable products then the all-important figure is the sum of the income from all the products. If this adds up to a total that exceeds the income of the primary product grown as a monoculture, then the farmer is better off, especially if the extra costs are minimal and the income comes at different times throughout the year. These benefits are multiplied if they also come with gains from environmental services, such as a lower susceptibility to pests and diseases or drought, or improved soil fertility – so involving reduced need for purchased inputs. This risk-aversion strategy is especially important if you are a smallholder living outside the cash economy and having to be self-sufficient. This way of thinking is often seen to run counter to the concept of 'economies of scale', but actually here too scale is important for the delivery of maximum benefits from the diversity. It seems to me that economists need to develop some new concepts that embrace diversity.

Before moving on I want to mention the need for shade crops. We've seen that sometimes agroforestry is about creating a landscape mosaic and sometimes it involves intimate mixtures of trees with staple food crops. In both cases we actually need some crops for the shady niches under trees. The main staple food crops do not fit the bill because they are bred for production in monocultures in full sunlight. This is perhaps a situation that could be changed. An alternative option is to seek new crops for these niches.

Eru (*Gnetum africanum*) is a candidate for this role. Eru used to be common in the forests of South-west Cameroon, but because of its very nutritious leaves (rich in protein and minerals) it has been heavily overexploited and is now becoming rare in the wild. It's not too hard to see why. During a visit to Cameroon I went to Idenau to see the export of eru to Nigeria. This was interesting as I had often seen vans taking eru to the border post with Nigeria. Each of these vans was absolutely stuffed solid with eru leaf and, in addition, had eru piled high on the roof, with at least as much on the roof as there was inside the van. In Idenau market we heard that 35 of these overloaded vans delivered leaf to the market every day. This highly exploitative trade has cleared the forest in South-west Cameroon and so the traders are moving further and further east to meet the demand from Nigeria.

As a vine, eru is a very suitable understorey crop in agroforests. After the early work by Patrick Shiembo to develop methods of rooting stem cuttings, the domestication of this species was taken up by staff of the Limbe Botanic Garden[4] in Buea, on the Cameroon coast at the foot of Mount Cameroon. Over the years, I have made several visits to these interesting gardens to see their work. Joseph Nkefor was doing experiments to identify the optimum light environment for production both under artificial shadecloth and in an agroforestry system with the eru growing up *Gliricidia* trees. This work has identified the best conditions and attained very good growth rates in these man-made systems. This means that there is now an excellent opportunity to develop eru as a new shade-adapted crop, so relieving the pressures on wild plants and allowing farmers to take advantage of the demand for export to Nigeria.

Earlier, I said that we have enough experience and understanding of agroforestry to start its widespread scaling-up. While I believe this to be true, there is still a great need for future research, especially in agroecology. As we have seen modern science has hardly made a scratch on the surface of this enormous puzzle. Modern science needs to wake up to the fact that our planet needs trees and we need to learn how they contribute to the vital ecological functions of the planet. I'm not going to spell out all the different research areas, as that would not be very interesting to most readers, but, if you wish to get a flavour of this you can look at the research agenda of organizations like ICRAF.[5]

Maybe this is an appropriate point to say something about the institutionalization of agroforestry as a science. ICRAF has about 70 internationally recruited research staff out of a total of about 330. In addition, there are national agroforestry research institutes in the USA and India. There are also agroforestry research programmes in many national agricultural research institutes around the world. This growing body of agroforestry research is perhaps illustrated by the attendance of the World Agroforestry Congress in 2009 by 1200 agroforestry scientists from 96 countries.

Agroforestry science has a dedicated, peer-reviewed, international science journal, *Agroforestry Systems*. It is published nine times a year by Springer. Agroforestry science is also published in many other scientific journals, including some of the top-rated ones. In addition, many academic books have been published about agroforestry. Together these research outputs and books are a resource used by about 229 tertiary colleges and universities delivering full degrees in agroforestry or agroforestry courses within other degrees in both developed and developing countries (Table 4.1).

Finally, in this chapter we have looked at how agroforestry provides agroecological functions in addition to ecological services and discussed the implications for environmental sustainability. Now we can move on in the next few chapters to see how we can intensify these ecologically sustainable farming systems, making them both more profitable and more sustainable socially and economically. We will see that intensification does not need to come with the price tag of environmental degradation.

Table 4.1. Tertiary colleges and universities teaching agroforestry to undergraduate and postgraduate students. (Source: World Agroforestry Centre – personal communication.)

Africa	Asia	Latin America	Developed countries	Oceania
Benin – 1	Indonesia – 6	Bolivia – 1	Australia – 3	Fiji – 1
Botswana – 1	Laos – 4	Brazil – 4	Europe – 15	
Burkina Faso – 3	Malaysia – 1	Colombia – 5	New Zealand – 2	
Cameroon – 5	Phillipines – 34 (9[a])	Costa Rica – 1	North America – 20	
Côte d'Ivoire – 1	South Asia – 30	Ecuador – 1		
Democratic Republic of Congo – 1	Thailand – 4	Guyana – 1		
Ethiopia – 6	Vietnam – 5 (1[a])	Honduras – 1		
Gambia – 1		Mexico – 2		
Ghana – 4		Nicaragua – 9 (3[a])		
Kenya – 6		Peru – 2		
Lesotho – 1		Venezuela – 1		
Liberia – 1				
Malawi – 3				
Mali – 2				
Mozambique – 2				
Namibia – 2				
Niger – 2				
Nigeria – 6				
Rwanda – 2				
Senegal – 3				
Sierra Leone – 1				
South Africa – 2				
Sudan – 5				
Swaziland – 1				
Tanzania – 5				
Togo – 1				
Uganda – 3				
Zambia – 2				
Zimbabwe – 3				

[a]Offering a full Bachelor of Science degree in Agroforestry.

Notes

[1] Unfortunately, shade-tolerant crops are few and far between, especially as most food crops are deliberately bred for cultivation in full sunlight. If agroforestry becomes more common in mainstream agriculture, it will be necessary for crops to be bred for shade environments.

[2] World Agroforestry Centre (2007) Agroforestry science at the heart of the three environmental conventions. In: *Tackling Global Challenges Through Agroforestry*. Annual Report 2006. World Agroforestry Centre, Nairobi, Kenya.

[3] Salati, E. and Vose, P.B. (1984) Amazon basin: a system in equilibrium. *Science* 225, 129–138.

[4] Joseph Nkefor, Nouhou Ndam and James Ackworth.

[5] www.worldagroforestrycentre.org

Further Reading

Beer, J., Ibrahim, M., Somarriba, E., Barrance, A. and Leakey, R.R.B. (2003) Establecimento y manejo de áboles en sistemas agroforestales. In: Cordero, J. and Boshier, D.H. (eds) *Árboles de Centroamerica: Un Manual para Extensionistas*. CATIE, Turrialba, Costa Rica and Oxford Forestry Institute, UK, pp. 197–242.

Ewel, J.J., O'Dowd, D.J., Bergelson, J., Daeler, C.C., D'Antonio, C.M., *et al.* (1999) Deliberate introductions of species: research needs. *BioScience* 49, 619–630.

Leakey, R.R.B. (1996) Definition of agroforestry revisited. *Agroforestry Today* 8, 5–7.

Leakey, R.R.B. (1999a) The use of biodiversity and implications for agroforestry. In: Leihner, D.E. and Mitschien, T.A. (eds) *A Third Millennium for Humanity? The Search for Paths of Sustainable Development*. Peter Lang, Frankfurt, Germany, pp. 43–58.

Leakey, R.R.B. (1999b) The evolution of agroforestry systems. In: Martin, G.J., Agama, A.L. and Leakey, R.R.B. (eds) *Cultivating Trees*. People and Plants Handbook No. 5. UNESCO, Paris, pp. 1–2.

Leakey, R.R.B. (1999c) Agroforestry for biodiversity in farming systems. In: Collins, W.W. and Qualset, C.O. (eds) *Biodiversity in Agroecosystems*. CRC Press, New York, pp. 127–145.

Leakey, R.R.B. and Tchoundjeu, Z. (2001) Diversification of tree crops: domestication of companion crops for poverty reduction and environmental services. *Experimental Agriculture* 37, 279–296.

Finding the Trees of Life 5

> Throughout the world agriculture is the site of a great diversity of managed and collected plant and animal foods. Conventional agricultural research and extension, by focussing only on the main food crops, chiefly cereals, roots and domesticated livestock, has long ignored the range of other plants and animals that make up agricultural systems.
>
> Ian Scoones, Mary Melnyk and Jules Pretty (1992) *The Hidden Harvest*. International Institute for Environment and Development, London.

> Domestication is about a whole lot more than fat tubers and docile sheep: the off-spring of the ancient marriage of plants and people are far stranger and more marvellous than we realize.
>
> Michael Pollan (2001) *The Botany of Desire: a Plant's-eye View of the World*. Random House, New York.

I have called this book *Living with the Trees of Life* in recognition of the thousands of species with a multitude of uses that are important to millions of people. The name also reflects how these species have impacted on my own life through my work to domesticate them.

Two billion people worldwide suffer from under nutrition and malnutrition: a combination of a lack of calories for energy and a lack of micronutrients for health. The agricultural Green Revolution focused on the domestication of staple starch crops like cereals, roots and tubers for the production of calories in a short growing season. Globally, much less attention has been paid to the wide range of fruits and vegetables that are rich in micronutrients. Despite our familiarity with domestication and the enormous benefits that can flow from it, we have actually applied it to surprisingly few species. For example, out of some 250,000 higher plant species, 20,000 of which have edible parts, we have only domesticated a little over 100 food plants. Some 30 other species have also been domesticated for other nonfood products, for example rubber and timber. On the other hand, we cultivate about 15,000 species as garden ornamentals. Does this suggest that we think aesthetics are more important than food?

Back in the first chapter I briefly described my involvement in the domestication of tropical timber trees and the package of techniques and strategies that we developed in Edinburgh. Then there was the flash of enlightenment in Kumba market (Fig. 1.1) with Patrick Shiembo when I saw the potential for implementing the package to domesticate the traditionally important food and medicinal trees. This led to 'Tropical Trees: the Potential for Domestication and the Rebuilding of Forest Resources' conference that I organized with Dr Adrian Newton in Edinburgh in 1992. The inspiration behind the conference was that people in developing countries are highly interested in their indigenous fruits and nuts, and aware of the characteristics that make them attractive. Thus there are many 'Cinderella' tree species that, if brought into cultivation, could diversify production, people's diets, the rural economy and farming systems, as well as having ecological roles in cropping systems.

One of the outcomes of the Edinburgh conference was that the domestication of these useful species should encompass socio-economic activities such as consultation with farmers about which species should be domesticated and linking this information with ethnobotanical knowledge so that it relates to the tradition and culture of the local people (Fig. 5.1). The concept that emerged also embraced the adoption of genetic resources by farmers and their management in production systems as part of sustainable land use practices to rebuild tropical forest resources.

At the interview for my job at ICRAF I described the techniques, strategies and opportunities to domesticate indigenous trees for non-timber forest products, as this was an area of work that was almost entirely absent from ICRAF's research programme at that time. This strategy paid off and I won my position at ICRAF on a tree domestication ticket. Consequently, on my arrival in Nairobi, I wanted to go ahead and implement a research programme aimed at lifting poor households out of the trap of subsistence farming by creating new cash crops that meet the needs of poor people. This approach, I believed, would add value to agroforestry. The problem, however, is the lack of uniformity and the variable quality of wild tree products, as well as the inconsistency of supply. This deters market development and wider commercialization. To overcome these problems we need to improve these species through genetic selection for improved quality, yield and regular production.

One of my early tasks was to raise funds for a tree domestication programme and I was fortunate to get the chance to address a gathering of the world's donors at CGIAR's Centres Week, in the offices of the International Monetary Fund in Washington, DC. I wanted to inspire them to be the fairy godmothers that would get 'Cinderella' to the ball. This worked and now many 'Cinderella' species are being domesticated all around the world,

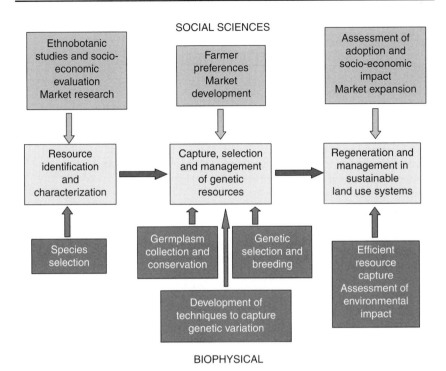

Fig. 5.1. Diagrammatic representation of the processes involved in the domestication of agroforestry trees.

thanks to the efforts of a growing number of people (Table 5.1). Before looking at this in more detail, let's consider the potential and why it is relevant.

The concept of 'domesticating' trees had been given little prominence before the Edinburgh conference, although of course we know it has happened because we recognize a few trees as horticultural crops (apples, citrus, plums, etc.), with selection, propagation and some breeding going back thousands of years. Some of these crops originate and are cultivated in the tropics (tea, coffee, cocoa, oil palm, etc.) and, although there has been some selection, the domestication process has not been very intensive. The same is true for timber trees. The problem is that trees are slow-growing, long-lived and have a long generation time. As a result, tree breeding is slow, laborious and unrewarding.

Compared with today's conventional food crops, traditional foods, which used to be gathered as common property (i.e. free) from forests and woodlands, are rich in nutrients (Table 5.2). These wild foods are important for a balanced, nutritious and healthy diet, as well as for medicines and many other products of day-to-day importance. Indeed, in the absence of social services and the cash to visit chemists, supermarkets and DIY stores, they

Table 5.1. Some of the tree species being domesticated that have potential as components of agroforestry systems.

Species	Use
Irvingia gabonensis/*Irvingia wombulu*	Kernels and fruits
Dacryodes edulis	Fruits and oils
Prunus africana	Bark for medicinal products
Pausinystalia johimbe	Bark for medicinal products
Ricinodendron heudelotii	Kernels
Gnetum africanum	Leafy vegetable
Barringtonia procera	Nuts
Inocarpus fagifer	Nuts
Santalum austrocaledonicum/ *Santalum lanceolatum*	Essential oils
Canarium indicum	Nuts
Sclerocarya birrea	Fruits and nuts
Triplochiton scleroxylon	Timber
Chlorophora excelsa	Timber
Swietenia macrophylla/*Swietenia mahogani*	Timber
Allanblackia species	Oils

Table 5.2. A comparison between the nutritional value (%) of the two top priority indigenous fruits/nuts in West Africa and three staple food crops.

Macronutrient	*Dacryodes edulis* fruit (88% DM)[a]	*Irvingia gabonensis* kernels (88% DM)	Maize grain (86% DM)	Rice grain	Cassava tuber (30–35% DM)
Carbohydrates	14	26–39	66–76	46–59	24–31
Fats/oils	32	51–72	2–6	1–2	<1
Protein	26	7.4	5–14	4–8	1
Fibre	18	1	1–3	1–4	1–2

[a] DM, dry matter.

are the local 'life-support' system of millions, if not billions, of poor and vulnerable people. Thus, the dominant focus of food-crop domestication on starch crops is seriously negligent, and perhaps even a breach of human rights, when the traditional food species are being lost as a result of deforestation for agriculture and urban development. The economic, social and cultural importance of these products should not be dismissed as trivial by agriculture. Modern agriculture has imposed an unhealthy diet on the poor, who basically now have to live outside the cash economy with often only poor access to their common-property supply of traditional foods. In this respect, modern agriculture fails to meet the needs of many farming households and is a feeble substitute for the traditional way of life.

Interestingly, for very different reasons, an unhealthy diet has led to obesity in rich countries and in the richer communities of developing countries. Nutritious food is important for both those who are under-nourished and those who are over-nourished. A more deliberate focus on crops rich in micronutrients is therefore important worldwide. When these micronutrients come from fruits and nuts, there are the added ecological and environmental benefits from the inclusion of trees in farming systems, which we discussed in Chapter 4. Thus crop diversification through agroforestry seems to be a much more logical approach to enriching our diets than trying to introduce genes to enrich the nutritional value of staple food crops – a process called bio-fortification.

Poor nutrition leads to a lack of immunity to disease. This is one of the factors that increase susceptibility to HIV/AIDS and reduce the effectiveness of anti-retroviral drug treatments. Consequently, the cultivation of indigenous fruits and nuts in agroforestry has considerable potential to reduce malnutrition and to stimulate trade, especially in local and regional markets.

Throughout the tropics there are thousands of these indigenous species that could be domesticated and brought into cultivation through agroforestry. Many of them are traded locally for a wide variety of traditional uses (e.g. Fig. 1.1). For example in Ghana well over 2000 species produce useful products. To illustrate this let's look at an example – *Garcinia kola*. Its fibrous twigs have bactericidal properties and are chewed as an alternative to toothbrush and toothpaste, while its bitter nuts (Fig. 5.2) are eaten as a stimulant and used to treat bronchial/chest problems, colic and headache. The flesh of the astringent fruit is a purgative (I can vouch for that!) and used as an anthelmintic to expel worms. The bark is used for tanning leather and to treat tumours, while the gum treats gonorrhoea and the latex is applied to cuts and wounds. The sap treats skin diseases

Fig. 5.2. Bitter kola (*Garcinia kola*) nuts (samples collected from different trees) in Cameroon.

and parasites, like ringworm. This list of uses for a single tree species is by no means exceptional.

On many occasions when I have been talking to poor farmers and asked them what they would like to grow on their farms, they look surprised. A common response is: 'Are you asking our advice – and not telling us what we should do?' So, in thinking about the future of tropical agriculture, I think we should be much more observant about what developing-country farmers say they want, and in some cases what they are actually doing. These farmers don't have a voice in setting the development agenda – what they do for themselves is the only way they can illustrate their needs and their actions speak louder than words. In this connection, it is clear from my experience that one of the things they are doing is expressing continued interest in, and need for, the indigenous trees producing fruits and nuts – the species that I am calling the Trees of Life.

So what is domestication? Probably the most widely accepted definition is that it is 'human-induced change in the genetics of a plant to conform to human desires and agroecosystems'. Other definitions emphasize 'pushing nature into a higher gear' and 'bringing into cultivation' and 'exposure to some form of management'. These definitions make it pretty clear that domestication is a deliberate act to improve the benefits for mankind derived from a useful species. The domestication literature recognizes that it can be done with differing levels of intensity. At the low end of the scale there is commensal domestication, in which a species becomes affiliated to the needs of mankind by their close proximity. Cats and dogs are thought to have been domesticated in this way originally. At the other end of the scale we have the very deliberate cross-breeding aimed at the development of particular genetic traits. This is the approach that has been used in agricultural crops and livestock, garden flowers and, more recently, also in cats and dogs.

Historically there have probably been two drivers of domestication. The first has been to meet 'demand' when it has been exceeded by 'supply'. The second is to create an improved product: better in yield, quality or appearance. In the case of agroforestry trees, the drivers have not been well recognized, as the demand is mainly from poor farmers in tropical countries who do not have a voice in the corridors of power, and who have limited ability to initiate the process for themselves. However, as we saw in the last chapter commensal domestication has occurred in Indonesia as farmers have developed their agroforests. We will see other evidence of this later.

How do we take advantage of the potential for tree domestication? To most people 'a tree is a tree is a tree' because it is not easy to recognize genetic variation, although it is staring you in the face! To understand the genetic variation in trees, I am going to use an analogy. The genetics of trees are actually very similar to the genetics of people, with variables that we can attribute to race, tribe, family likenesses and individuals. Think of a 'Miss World' or 'Mr Universe' competition(!) – or even the selection of a marriage partner or mate. The principles are the same. In a beauty competition, the judges are looking for contestants that combine good physical attributes (I guess you can think of a list of body parts that vary in size and, depending on what turns you on, you can select them in order of importance), as well as things such as hair colour, skin colour, eye colour, etc. These things are generally measurable. Then the judges also consider characteristics like personality, intelligence, etc., which are more difficult to quantify. The judges' final selection is the contestant who, on average, appeals to the different panel members the most. Not everyone is going to agree with the choice of winner, because people differ in what they find attractive.

How does this compare with identifying superior trees? Well, it is exactly the same. Even within a single village we will find that every tree is measurably different from every other in things like the size of their fruits and nuts, the number and size of branches, the straightness of the trunk, etc. – these are the physical traits. Then there is variability in the colour of skin and pulp of the fruits and in the kernels of the nuts, as well as in the leaves, the bark and the wood. The fruits and nuts will also vary in flavour (sweetness, bitterness, etc.) and many other ways – these are the quality traits. Finally there are many other ways in which the trees will vary…things like the seasonality of production. Some of these traits are much more difficult to assess, such as the nutritional value, medicinal value and chemical content of their products.

When talking to farmers about genetic variability we discuss how they are different from other tribes and races before going on to consider whether or not their children are identical copies of mother or father, or indeed identical copies of their brothers and sisters. We obviously end up recognizing that everyone is different from everyone else, but that it is possible to see family, tribal and racial patterns. So it is with trees, although you may never have thought about trees in this way. Next time you look at two similar-sized trees of the same species growing side by side, see if you can spot these differences. Sometimes they can be strikingly different in branching pattern, for example. How do we know that these differences are due to the genetic variation between individuals? Well, that's not possible by just looking at them, but it's easy to prove if you are willing to take time and effort. All you have to do is to take ten cuttings from each of ten trees, root them, and plant the cuttings from each tree at the same site as a different line. Then, as the trees grow, you will be able to see that the lines of each tree are recognizably different from each other, but that the trees within each row are remarkably similar. This is because they are genetically identical, while the trees of the different rows are genetically different.

Let's move on from beauty shows with people to dog shows like Crufts. In the latter, we select the best individuals within different breeds – but they are all the same species derived from the wolf. Over a very long period of time what has happened during the domestication of dogs is that the breeders have selected specific traits found within the overall genetic variability of the wolf, teased them out and then 'concentrated' them by breeding 'like' with 'like'. This illustrates that the genetic code of the wild wolf contains all the variation that we now see in the hugely variable breeds of the domesticated dog.

How do we know that trees are so variable? Well, we have measured them, as we will see in more detail later. For now, the point is that trees are just as variable as people and in very much the same ways as people. Like wolves, individual tree species also contain the variability we now see expressed in dog breeds. This is because during reproduction, the genes of the mothers and fathers of trees, people and dogs are distributed at random during fertilization – this is called outbreeding. This is what gives us the enormous opportunity to domesticate trees through the selection of the superior individuals – what we call 'plus-tree' selection. These 'plus-trees' can then enter a breeding programme or, as we will see later, they can be propagated vegetatively by taking cuttings or grafting.

In trees, vegetative propagation gives us a powerful technique we can use to capture and copy the genetic traits of any individual tree as rooted cuttings, or as grafts. In this way we can capture the rare individuals in the wild population that have specific combinations of desirable traits that meet the needs of different markets and new industries – the plus-trees. Then using the vegetative propagation techniques again, these plus-trees can be massed produced as genetic copies of the original tree, creating a cultivar (cultivated variety) in large quantities. In this way, we can make crop varieties that deliver a quantum leap in quality and yield improvement in just a few years. That quantum leap could then be further expanded many, many times over a longer time scale by controlled breeding between selected cultivars, giving rise to tree cultivars that are as distinct as dog breeds and designed to meet all sorts of market opportunities. Thus the tree crops of the future could be very different from today's wild species, just as modern dog breeds differ from their ancestral progenitor, the wolf.

So, there are many wild species with the potential to become new crops. Each of these is likely to have genetic variation in traits that confer improved marketability, and there is a big demand for their products in the marketplace. That is all very encouraging. Now, before considering how to do the job, let's think about an implementation strategy.

Conventionally, crop breeding is the prerogative of national and international research institutes, biotechnology companies and universities, typically implemented by geneticists working with little interaction with the farmers until a variety has been produced and is ready for field testing. This strategy is expensive and the farmer gets little benefit until the lengthy process of doing experiments, data analysis and writing research papers has been completed. There were four reasons why this strategy seems to be inappropriate for agroforestry trees: (i) the research budget is small; (ii) the results are needed quickly; (iii) the traditional knowledge of the farmers could help to give appropriate direction to the research; and (iv) the impacts from small inputs to many species would result in greater socio-economic impacts than an intensive input to one or two species.

For all four reasons it seems it would be more beneficial to do this work on farms and within the communities rather than in research stations. This would have the added benefit that, as the farmers gained experience with the techniques, they could adapt and apply them to other species. In other words, this would be a participatory, self-help process in the community done by the community – so that the villagers are the beneficiaries of their own actions. Such an approach should empower these communities and hopefully give them the incentive to improve their own living standards and to lift themselves out of poverty. Similar approaches to other aspects of rural development in developing countries have been found beneficial for promoting their adoption by farmers. This approach to 'development' through the sustainable use and improvement of genetic resources, is in accordance with the international Convention on Biological Diversity of 1993, an outcome of the Earth Summit in Rio de Janeiro in 1992.

The consultative and participatory process started with a discussion of species priorities, based on which forest-tree species the farmers would like to see domesticated as new crops; a long list emerged. The farmers in Cameroon and Nigeria had plenty of ideas about what they would like to grow, mostly tree species virtually unknown to science, and certainly unknown to most people in the rich countries. Typically, farmers identified indigenous fruits and nuts as the most important species.

The species that was their top priority was bush mango (*Irvingia gabonensis* – Fig. 1.2). It is a large greenish-yellow fruit, with somewhat stringy yellow flesh around a large woody seed – so it looks like the domesticated mango, but is actually unrelated. The flesh is not very sweet and the main reason for its popularity is the kernel inside the nut. These kernels are dried, stored and used as a food-thickening agent in soups and stews as they secrete a much appreciated mucilaginous slime.

Priority number two was safou (Fig. 5.3) – also known as African plum or African pear (*Dacryodes edulis*). The purple-blue fruits of safou are very common in all markets during the fruiting season (May–August). They are typically eaten roasted with a main meal, more as a vegetable than as a fruit. The fruits are oily and highly nutritious (Table 5.2), with a nutty flavour, tasting something like a cross between an avocado and an olive. Personally, I think they are very tasty stuffed with minced meat and eaten with a salad. Safou trees are commonly planted in compound gardens and as the shade trees for cocoa and coffee in Cameroon and make up 20–60% of all fruit trees in farmers' fields. They play an important part in the economy of rural

Fig. 5.3. *Dacryodes edulis* (safou) tree in Cameroon.

communities, especially as the income from the sale of fruits buffers the fluctuations of international prices of cocoa and coffee.

The third priority for Cameroon was njangsang (*Ricinodendron heudelotii* – Fig. 1.3). This is a fast-growing pioneer tree, which invades cleared areas such as fallows. It would probably be much more common in farmers' fields if the seedlings were not very vulnerable to attack by psylids (woolly aphids), which causes the leaves to wrinkle up and fall off. The product that is consumed and marketed locally is not the fruit, but a golden, pea-sized kernel from the nut. It is used as a flavouring, which tastes a bit like groundnut (peanut), especially in fish dishes.

The fourth priority was bitter kola (*Garcinia kola* – Fig. 5.2). Its many uses have been mentioned already, but primarily the very bitter nuts are eaten raw, as a stimulant – an acquired taste. Apparently, it has also been used in beer making as an alternative to hops.

The fifth priority was the star apple (*Chrysophyllum albidum*). It is eaten as a fresh fruit and was an especially high priority in Nigeria.

It was interesting that when this process was repeated in other regions around the world, generally indigenous fruits and nuts were selected as the farmers' priority species (Table 5.3); Amazonia was the only exception. There farmers chose three species for their importance as producers of poles and wood. One of the other choices was peach palm, which in addition to producing tasty fruits is the source of heart of palm. Palm hearts are the soft central parts of the main stem of young sucker shoots. They are a delicacy that is widely eaten and traded in the Americas. This over-riding focus of farmers on edible fruit and nut species is, however, very interesting as it emphasizes a point that I have made

Table 5.3. Species priorities for domestication.

Region	Species	Product	Uses
Humid lowlands of West Africa	1. *Irvingia gabonensis* – bush mango	Kernel	Food thickening
	2. *Dacryodes edulis* – safou	Fruit	Food
	3. *Ricinodendron heudelotii* – njangsang	Kernel	Spice
	4. *Chrysophyllum albidum* – star apple	Fruit	Food
	5. *Garcinia kola* – bitter kola	Kernel	Stimulant
The Sahel	1. *Adansonia digitata* – baobab	Fruit/leaf	Food
	2. *Vitellaria paradoxa* – shea nut	Kernel	Food
	3. *Parkia biglobosa* – néré	Kernel	Food
	4. *Tamarindus indica* – tamarind	Fruit	Food
	5. *Zizyphus mauritiana* – ber	Fruit	Food
Southern Africa	1. *Uapaca kirkiana* – mahobohobo	Fruit	Food
	2. *Sclerocarya birrea* – marula	Fruit/kernel	Food
	3. *Zizyphus mauritiana* – ber	Fruit	Food
	4. *Vangueria infausta* – wild medlar	Fruit	Food
	5. *Azanza garckeana* – snot apple	Fruit	Food
Peruvian Amazonia	1. *Bactris gasipaes* – peach palm	Fruit/palm heart	Food
	2. *Cedrelinga catenaeformis* – tornillo	Stem	Wood
	3. *Inga edulis* – guaba	Fruit	Food
	4. *Calycophyllum spruceanum* – capriona	Stem	Wood
	5. *Guazuma crinite* – bolaina blanca	Stem	Wood
Indonesia	1. *Paraserianthes falcataria* – sengon	Stem	Wood
	2. *Artocarpus heterophyllus* – jackfruit	Fruit	Food
	3. *Aleurites molucana* – candlenut	Kernel	Food
	4. *Artocarpus altilis* – breadfruit	Fruit	Food
	5. *Parkia speciosa* – petai	Fruit	Food

several times – local people have in the past been highly dependent on these species and, in addition to knowing and being very familiar with them, they are also important components of local culture and tradition.

The species list developed during this process was then reduced to the top five per region (e.g. Fig. 5.4) by also considering the species commonly seen on the roadside and in urban markets. The importance of market forces was specifically integrated with the farmers' priorities because the success of domestication depends on commercial demand. In other words, this process has to be both 'farmer-driven' and 'market-led'.

When farmers were asked what characteristics they would like to see improved by genetic selection they, not surprisingly, highlighted fruit size and taste. Interestingly, they also commonly mentioned that they would like trees with shorter stems for easier picking and earlier fruiting. These then were the objectives that we pursued in Cameroon.[1] They were implemented by a team of ICRAF staff and partners from the Institut de la Recherche Agricole pour le Développement (IRAD).[2] This process was subsequently implemented in the Sahel region of West Africa, southern Africa, Latin America and South-east Asia.

The participatory tree domestication process by which rural communities select, propagate and manage trees according to their own needs involves a partnership with scientists. This process involves the integration of the farmers' own efforts with the conventional science pathway typically implemented on a research station with laboratory support. Participatory domestication, on the other hand, is usually implemented on-farm at the community/household level and oriented towards specific local markets and encompasses the use of both indigenous knowledge and genetic selection based on scientific principles. In our participatory approach to domestication, researchers typically act as mentors, helping and advising the farmers on technical issues, and sometimes jointly implementing on-farm research. One advantage of working directly with farmers is that the project outputs are immediately disseminated and adopted by the participating community, thus overcoming the delays that often arise from the research station pathway for the domestication process. The researchers can also be the link with the markets and with industry so that the development of cultivars is informed by an understanding of the needs of both conventional and emerging markets. Ultimately, however, it is the farmer who creates, multiplies and grows the cultivars that are produced and who reaps the benefits.

The process of participatory domestication involves many activities, all of which require oversight and mentoring by scientists and technicians. We will see later that this approach can also involve civic authorities, non-governmental organizations (NGOs) and commercial companies. The role of the scientists in this process is, first, to ensure that the farmers learn and correctly implement the techniques to select, capture and multiply the best

Fig. 5.4. Participatory priority setting in villages in Cameroon. (a) Farmers discuss their priorities; (b) kernels of bush mango (*Irvingia gabonensis*); (c) fruits of safou (*Dacryodes edulis*); (d) kernels of njangsang (*Ricinodendron heudelotii*); (e) bitter kola (*Garcinia kola*) nuts; (f) fruits of star apple (*Chrysophyllum albidum*).

genetic material. Secondly, the scientists are responsible for ensuring that farmers have appropriate understanding of the techniques and strategies. This strategy and process is equally appropriate for the domestication of tree species producing: (i) fruits and nuts; (ii) medicinal products; (iii) leafy vegetable and animal fodder; (iv) timber and wood; and (v) extractives like essential oils, resins, etc.

With the emphasis of participatory domestication on village-level activities, it is clear that to be adoptable the techniques have to be simple, robust and inexpensive. Consequently, the vegetative propagation techniques, for example, have been developed to be appropriate for village nurseries in remote locations without piped water or electricity.

One of the commonly raised concerns about domestication is that the selection processes lead to a loss of genetic diversity, as has happened in some highly bred crops, and indeed within any particular breed of dog. This is most common when domestication has occurred over millennia, with minimal supervision. In some extreme cases the domesticated individuals lose the ability to cross with their wild derivatives. Care should be taken to reduce this risk. For example, some wild populations should be conserved for future use in genebanks.

In the case of indigenous fruit trees, the current level of genetic diversity being lost is minimal and there are still large wild populations in adjacent areas. Furthermore, the village-level selection programme means that the diversity attributable to different villages is not being lost as each village develops its own, and very different collection of cultivars. We should also acknowledge that an outcome of a wise domestication strategy is also a means of conserving genetic resources, especially in species that are being over-exploited in the wild. In a managed programme of domestication, serious loss of genetic diversity should only occur under circumstances of severe mismanagement or the abuse of the domestication strategy.

As already stated, through the promotion of cultivar development at the village-level, participatory domestication allows farmers to be the beneficiaries and guardians of the use of their traditional knowledge about different species, the within species variation and the planting stock derived from it. Although the rights are enshrined within the Convention on Biological Diversity[3] and the World Trade Organization (WTO) Agreement on Trade-Related Aspects of Intellectual Property Rights (TRIPS), the international legal instruments to enforce them are, at best, weak. Thus, unfortunately, there is still some considerable way to go in the development of internationally binding legal instruments to protect the intellectual property of poor people from exploitation by unscrupulous entrepreneurs.

Through the community-based implementation strategy of participatory domestication, we strive to provide an example of best practice – a politically acceptable approach to bioprospecting/biodiscovery, which is the antithesis of 'biopiracy'. However, farmers engaged in participatory domestication need to be fully informed of the issues so they understand their rights and know how to maintain and protect themselves. Important areas for resolution include the complex issues of access and benefit sharing with regard to the commercialization of genetic resources. Currently, interim procedures are being developed by ICRAF to minimize the risks of farmer exploitation. Hopefully, these will be compatible with any future international laws.

After I had left ICRAF and returned to CEH Edinburgh, I was involved in a DFID-funded project to gain a better understanding of the relevance of agroforestry tree domestication to subsistence farmers in Cameroon and Nigeria. This involved two components: (i) to quantify the tree-to-tree variation in fruit traits; and (ii) to examine the socio-economic constraints and benefits of bringing indigenous trees into cultivation, this component being relevant to this chapter. The latter was led by Dr Kate Schreckenberg of the Overseas Development Institute in London. This study involved participatory community-level research, household surveys and whole-farm fruit-tree inventories. It concluded that farmers in the study area were indeed very interested in the cultivation of indigenous fruits and that the approach was very relevant to their needs.

Safou was found to be very important for home consumption, but additionally its fruits were an important source of income for women. A crop of safou fruits can be worth between US$20 and US$150 per tree, depending on the quality of the fruits and the yield. In two of the four communities surveyed, safou was ranked higher than all other tree species for its food value. Women particularly like the fact that the boiled or roasted fruit can be eaten with cassava, providing a meal that is quick and easy to prepare at a time when most of their labour has to be devoted to agricultural activities. Additionally, this income stream comes at a time of year when school fees are due and other sources of income were scarce. So improving the income of the mothers from safou is likely to lead to better access to education for the children.

In some areas land tenure is a constraint to the planting and management of trees, however in southern Cameroon land ownership was not found to be a problem as most households have at least some land with secure tenure. Labour availability was not a particular problem either, as tree-planting and maintenance work could be integrated with that of other tree crops, and harvesting was either done by family members or by wholesalers who bought the crop on the trees and did their own harvesting.

I have mentioned medicinal tree species several times, but so far not said much about them. The bark of two species found in Cameroon is exported to pharmaceutical companies in Europe, as well as being important in traditional medicine. The first is pygeum (*Prunus africana*), which I mentioned earlier was also important in Kenya, and the second is yohimbe (*Pausinystalia johimbe*). As both species are being over-exploited in the wild, it was decided to add them to the domestication programme.

Pygeum is widely distributed in sub-Saharan Africa, occurring in montane areas over 1200 m and with annual rainfall above 1000 mm. These biophysical requirements mean that the species occurs as distinct 'island' populations, a situation typically associated with genetic variation between 'islands'. These montane forest areas are also important for endangered endemic birds (Bannerman's turaco) and monkeys (Preuss's gueron), which eat the fruits of pygeum. However, it is the bark of the tree that is of pharmaceutical interest and is used to treat benign prostatic hyperplasia in elderly men, making it a valuable product on international markets. Incidentally, when visiting western Uganda, near the Bwindi National Park, I remember discovering that elephants also appreciated the bark. I presume male elephants have a prostate gland too and maybe they understand the benefits of eating the bark!

Commercially, pygeum bark can be harvested sustainably, but at this time unsustainable harvesting or even the felling of the whole tree is common practice, raising concerns about the future of the species in both commercial companies and conservation organizations. Blood pressures were running high in the mid-1990s on all sides of the debate, and so we decided to initiate a domestication programme in partnership with Limbe Botanic Garden. The strategy and approach of this project basically followed the general pattern described earlier in this chapter so that smallholder farmers could be the beneficiaries of its medicinal and social use.

Yohimbe is another wild tree that grows in the lowland forests of southern Cameroon. Like pygeum, it too is over-exploited, this time because the bark extracts are nature's equivalent of 'Viagra' and are used to treat impotence in men. Yohimbine, the main ingredient, is also a heart stimulant used in the treatment of a number of cardiac disorders. An overdose can have dangerous side effects. Its reputation as an aphrodisiac accounts for its inclusion in many products marketed by sex shops in developed countries. It is also used by body builders to promote fat loss.

The international market for aphrodisiacs is substantial and many of them are based on ingredients of dubious efficacy, such as rhino horn and tiger body parts. If yohimbe extracts could be an alternative to these unacceptable products from threatened wildlife, then farmers could be the beneficiaries, while simultaneously reducing the pressures on wildlife populations. If you are thinking that this agroforestry outcome could have negative consequences for population growth, then there is another woody plant that could be grown in agroforestry systems – the thunder god vine (*Tripterygium wilfordii*) found in the Far East. Among its many uses is that of a contraceptive, due to its capacity to reduce sperm motility.

Finally, before we go on to examine the techniques to implement agroforestry tree domestication in the next chapter, I am going to introduce the term agroforestry tree products (AFTPs). I introduced this term to the scientific literature in 2004 with Tony Simons, at the 25th Anniversary of ICRAF. The idea behind creating a new term for the wide range of products that can be derived from trees grown in farmland was to distinguish them from the products from wild forest trees. When gathered from natural forests the products are typically called 'non-timber forest products' or 'non-wood forest products'. These are what are known as common property extractive resources; in other words, they do not belong to anyone in particular and can be gathered in the forest and used by anyone. The importance of these extractive resources was recognized when people promoting forest conservation were trying to demonstrate the value of intact forest. This importantly showed that forests do have a monetary value in excess of the timber standing in them. However, as we have seen, farmers are deliberately growing forest tree species on their farms for a range of tree products. So our newly named AFTPs are not common property extractive resources, but are in fact the products from new tree crops belonging to the farmer on whose land they have been grown. Statistics on the use of tree products of relevance to the development of rural development policies were, therefore, overvaluing forest production and undervaluing agricultural production. This inaccuracy was additionally reducing the perceived value of agroforestry, something that affects the importance placed on agroforestry by national governments, and indeed by international donors funding development projects.

In the next chapter we will see how to convert wild trees producing marketable products into new crops with the ideal combination of characters specific to particular uses and industries.

Notes

[1] Drs Steve Franzel, Willem Janssen and Hannah Jaenicke, Doug Boland and Julia Ndungu.
[2] Drs Bahiru Duguma, David Ladipo and Zac Tchoundjeu and Joseph Kengue, supported by Ann Degrande, Joseph Kengue, Jean-Marie Fondoun and others.
[3] Recently expanded within the Nagoya Protocol (2010).

Further Reading

Akinnifesi, F.K., Leakey, R.R.B., Ajayi, O.C., Sileshi, G., Tchoundjeu, Z., *et al.* (2008) *Indigenous Fruit Trees in the Tropics: Domestication, Utilization and Commercialization.* CAB International, Wallingford, UK.

Boshier, D.H., Mesén, F., Hughes, C. and Leakey, R.R.B. (2003) El valor de la diversidad y la calidad de la semilla en la plantación de árboles. In: Cordero, J. and Boshier, D.H. (eds) *Árboles de Centroamerica: un Manual para Extensionistas.*

Centro Agronómico Tropical de Investigación y Enseñanza (CATIE), Turrialba, Costa Rica and Oxford Forestry Institute, Oxford, pp. 283–302.

Leakey, R.R.B. (1997) Domestication potential of *Prunus africana* ('pygeum') in sub-Saharan Africa. In: Kinyua, A.M., Kofi-Tsekpo, W.M. and Dangana, L.B. (eds) *Conservation and Utilization of Medicinal Plants and Wild Relatives of Food Crops.* United Nations Educational, Scientific and Cultural Organization (UNESCO), Nairobi, pp. 99–106.

Leakey, R.R.B. (1999) Potential for novel food products from agroforestry trees. *Food Chemistry* 64, 1–14.

Leakey, R.R.B. and Akinnifesi, F.K. (2008) Towards a domestication strategy for indigenous fruit trees in the tropics. In: Akinnifesi, F.K., Leakey, R.R.B., Ajayi, O.C., Sileshi, G., Tchoundjeu, Z., et al. (eds) *Indigenous Fruit Trees in the Tropics: Domestication, Utilization and Commercialization.* CAB International, Wallingford, UK, pp. 28–49.

Leakey, R.R.B. and Izac, A.-M. (1996) Linkages between domestication and commercialization of non-timber forest products: implications for agroforestry. In: Leakey, R.R.B., Temu, A.B., Melnyk, M. and Vantomme, P. (eds) *Domestication and Commercialization of Non-timber Forest Products in Agroforestry Systems.* Non-Wood Forest Products No. 9. Food and Agricultural Organization (FAO), Rome, pp. 1–7.

Leakey, R.R.B. and Newton, A.C. (1994a) *Tropical Trees: Potential for Domestication, Rebuilding Forest Resources.* HMSO, London.

Leakey, R.R.B. and Newton, A.C. (1994b) *Domestication of Tropical Trees for Timber and Non-timber Forest Products.* MAB (Man and Biosphere) Digest No. 17, United Nations Educational, Scientific and Cultural Organization (UNESCO), Paris.

Leakey, R.R.B. and Newton, A.C. (1994c) Domestication of 'Cinderella' species as the start of a woody-plant revolution. In: Leakey, R.R.B. and Newton, A.C. (eds) *Tropical Trees: the Potential for Domestication and the Rebuilding of Forest Resources.* HMSO, London, pp. 3–4.

Leakey, R.R.B. and Simons, A.J. (1998) The domestication and commercialization of indigenous trees in agroforestry for the alleviation of poverty. *Agroforestry Systems* 38, 165–176.

Leakey, R.R.B. and Tomich, T.P. (1999) Domestication of tropical trees: from biology to economics and policy. In: Buck, L.E., Lassoie, J.P. and Fernandes, E.C.M. (eds) *Agroforestry in Sustainable Ecosystems.* CRC Press/Lewis Publishers, New York, pp. 319–338.

Leakey, R.R.B., Temu, A.B., Melnyk, M. and Vantomme, P. (eds) (1996) *Domestication and Commercialization of Non-timber Forest Products in Agroforestry Systems.* Non-Wood Forest Products No. 9. Food and Agricultural Organization (FAO), Rome.

Leakey, R.R.B., Wilson, J. and Deans, J.D. (1999) Domestication of trees for agroforestry in drylands. *Annals of Arid Zones* 38, 195–220.

Leakey, R.R.B., Schreckenberg, K. and Tchoundjeu, Z. (2003) The participatory domestication of West African indigenous fruits. *International Forestry Review* 5, 338–347.

Schreckenberg, K., Leakey, R.R.B. and Kengue, J. (2002) A fruit tree with a future: *Dacryodes edulis* (safou, the African plum). *Forests, Trees and Livelihoods* 12, 1–147.

Simons, A.J. and Leakey, R.R.B. (2004) Tree domestication in tropical agroforestry. *Agroforestry Systems* 61, 167–181.

Selecting the Best Trees   6

> Food crop domestication has been the precursor of settled, politically centralized, socially stratified, economically complex and technologically innovative societies.
> Jared Diamond (1999) *Guns, Germs, and Steel: the Fates of Human Societies.* W.W. Norton & Co., New York.

> There is clearly immense scope for selecting genetically superior trees of all sorts and for all purposes … future growers of timber, fuel, fodder and food trees stand to profit from international forestry and agroforestry research that is now being initiated.
> Derek Tribe (1994) *Feeding and Greening the World: the Role of International Agricultural Research.* CAB International, Wallingford, UK.

We have just seen that there is enormous variability in trees that equates to the variability we see in human races, tribes, families and individuals. All of this variation affects how a tree grows, how it can be propagated and how it yields different products. Then, most dramatically, we see that the tree products themselves vary in shape, size, colour, flavour, chemical content, etc. This is both a great asset and a nightmare. The nightmares stem from the fact that this huge variability seems to be more or less distributed at random between trees, so it is difficult to unravel, categorize and use. However, if we can achieve this, the extent of this variation creates huge opportunities for selection.

So, how do we unravel all this diversity and identify the best trees – the 'plus-trees'? It is probably obvious that one way to do it is by measuring everything. This is perhaps the best way to do it, but it is not very practical. So we need to find other ways. The simplest way, when we are working with local people, is to let them make the selections as they are in fact very knowledgeable about the trees on and around their land. For example, in Cameroon, villagers will say things like: 'If you go down the left-hand path that crosses the river and then go up the hill to the big ayous tree, you will see a bush mango tree. It has sweet and tasty fruits, while the one on the other path, by the cocoa plot, has big kernels that fetch a good market price.' Likewise, in Namibia, it is obvious that the farmers know every tree because every tree has a pet name, often a name that describes the particular characteristic of that individual tree.

Very wisely, however, these farmers don't show you their best trees when you turn up in their village asking questions about their trees. In the early days of setting up participatory projects and engaging with villagers in the lead-up to starting a village nursery, we were often shown fairly average trees reasonably close to the homestead, rather than their best trees. We therefore used these trees to demonstrate how to vegetatively propagate a mature tree. This reticence to show us the best trees is actually good, as we usually have to lop off some branches and they would not like this happening to their best trees until they are convinced about the value of what we are doing. This is part of the process of building trust and a genuine partnership with the farmer and it takes time. It is for the scientists to prove their credentials and to prove themselves worthy of that trust. When the farmers see that the techniques work, and that the plants produced from them are true copies of the trees they came from, then they usually engage fully. I should reiterate here that the concept of participatory domestication is that we teach the farmers basic techniques. It is entirely up to them how they use the knowledge and which trees they propagate.

For us, as the scientists developing the knowledge and expertise to be used by the tree domestication teams mentoring the farmers, we had to gain a good understanding of the tree-to-tree variation in fruits and nuts. The idea behind my work in the DFID project in Cameroon was therefore to collect detailed quantitative data about the fruits and nuts of the two top priority species in Cameroon (safou – *Dacryodes edulis* – and bush mango – *Irvingia gabonensis*). From this we could learn about the extent of the variation in each trait and so gain an understanding of the potential to make significant improvements in fruit, nut and kernel sizes and quality, and subsequently how to match this to the needs of different markets and industries. Once we had this knowledge we could then feed it into the farmers' training courses. Up to this time, the variability in the fruits and nuts of these species had been described subjectively. To do our work we needed to visit different villages and to weigh and measure 24 fruits from each of 100 trees of each species in each population. We decided to go to four villages in southern Cameroon and four villages in south-west Nigeria.

Each freshly collected fruit was measured on site and then dissected into its component parts (flesh and nut, and then the nut into shell and edible kernel). I worked with two Masters students (Kijo Waruhui and Alain Atangana) in Cameroon and with Paul Anegbeh, who then worked with two students (Victoria Okafor and Cecilia Usoro) in Nigeria to develop the field techniques, and then the students did the hard work going around the villages with support from project staff.[1] They also made some subjective assessments of taste, oiliness and fibrosity for each fruit sample, as well as asking the farmers for their assessment of overall quality and the potential market price.

I enjoyed my small contribution to this field work. It was fascinating to be taken out into the farm and forest behind these homesteads. Before this project, I had visited and walked around many forest reserves in West Africa and so was fairly familiar with the forest and many of the tree species in it. However, I had never previously had the chance to go into these farms; you are not too sure what kind of response you, as a foreigner, will get from the owners and I never liked to impose on their lives. In this project we had a great excuse to visit the homes of many people and to talk to them about their trees and learn from them about their use of tree products.

We always got a great welcome and if the farmer was close to the house we were nearly always given a tour of at least the area around the homestead. These tours often went much deeper into the forest, particularly when we were with farmers participating in this project. Typically there is a well-walked path as you go away from the house, and as you go deeper into the mosaic of farm, old fallow plots and forest, so your path is joined by others from other houses. In addition the path often divides and crosses streams and swampy patches. The locals are interested to see how the foreigner will cope with crossing a stream or bog by walking along a log. Fortunately, I have never fallen off – yet. Commonly these paths go very close to the most important fruit-producing trees, so the day's work is also punctuated with opportunities to taste the fruits and leaves of different species.

These walks are an opportunity to illustrate to the farmer that, while I may not be a local, I am not completely ignorant of his/her world and that I have great respect for it. They are very sensibly cautious when visitors appear in their villages, especially when they are heralded as 'experts' for an aid project. One way of showing that you are knowledgeable is to name and ask questions about the uses of the species you recognize as you pass, and to comment on seeds and fruits lying beside the path. Sometimes, too, it is possible to get across your knowledge about local culture and tradition by, for example, commenting when you see a tree with the symbolism of a local 'ju-ju' (traditional magic charm) to warn off raiding parties! Such trees are usually the household's favourite trees in terms of flavour or fruit size. Much as I would have liked to photograph these 'ju-ju' symbols, I knew that it would cause offence. Indeed, the farmer might be concerned that I would show the photos to a witch doctor and ask him to break the spell. Anyway, by recognizing the signs, it was possible to both demonstrate my knowledge of and respect for local culture and have a discussion with the farmer about the principles and opportunities of tree domestication.

Another way to demonstrate that as a 'white guy' you are not a total idiot is to try to memorize the path junctions on the outward journey and then to lead the walk back to the farm. This can be a big challenge and, of course, I frequently go wrong or have to ask which path to take. While this is a small and perhaps silly game, it is my impression that it does boost the confidence of the farmer that you are not just a 'pen-pusher' from Europe, a misdirected 'expert' or someone there to steal their traditional knowledge for personal gain. From my point of view, gaining this first-hand experience and understanding of the farms and what interests the farmers is critically important for our task of trying to implement appropriate techniques and strategies to improve the lives of poor smallholder farmers.

Our project showed that the tree-to-tree variation in everything we measured had continuous variation. This means that if you arrange the data from the smallest fruit to the largest fruit in sequence, you find that the average fruit size per tree increases steadily (Fig. 6.1) and that there are no groupings of trees that have fruits of a similar size. This basically shows that there are no clusters of trees with similar sized fruits that we might call 'varieties', but rather that the population is made up of individuals expressing the kind of variation you would expect in a random sample. Occasionally, we also found that these samples included some unusually big fruits. Thus, although fruit weights in bush mango in both Cameroon and Nigeria were mainly in a similar range (50–200 g), there were a few trees with much bigger fruits in Nigeria (200–400 g). Likewise, in safou the results were the other way around, with fruits in both countries varying from 10 to 70 g, but in Cameroon a few were bigger (70–110 g).

A second interesting finding was that the nutshell thickness varied and, at the extreme, a few trees had fruits whose nuts were so thin-shelled that they were very easily broken to remove the kernel. This trait (described as

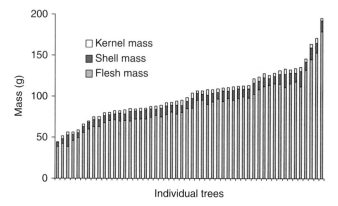

Fig. 6.1. The typical pattern of continuous variation in fruit size in village populations of indigenous fruit and nut trees.

'self-cracking') had been reported once before and was of special interest, as the cracking of bush mango nuts is very difficult and requires skilled use of a machete. If we could find trees that combined self-cracking with the presence of big kernels, then we could have cultivars from which it would be much easier to harvest the kernels and with a lower risk that the ladies using the machetes would cut their fingers.

When we analysed our data in more detail, we found our most important, but not unexpected, result. It was that high values of one trait are not necessarily associated with high values of another trait. For example, large fruits are not necessarily sweet fruits, and large fruits do not necessarily contain large nuts or kernels. This means that the more traits you measure, the greater the overall level of variability there is between trees. The magnitude of this variation is good news in terms of the opportunities for plus-tree selection, but the volume of data soon becomes mind-blowing. One additional and very important finding was that all this variation is recognized by the retail markets, so the fruits from different trees are sold at different prices as the consumers register their approval of certain combinations of different traits (Fig. 6.2).

You will remember that we were doing this exercise not to identify the trees for selection, as that is for the farmers to do. We were doing it to get an understanding of the variation, so that we could provide guidance to both

Fig. 6.2. Variation in fruit size and colour and in market price of *Dacryodes edulis* (safou) fruits from different trees in Cameroon.

farmers and perhaps buyers from different industries. Consequently, the practical approach is to seek trees that have particular, market-oriented, trait combinations – such as big, sweet fruits for the fresh-fruit market, or big, easily extracted, oil-rich kernels for the kernel market. To pursue this approach we developed what are called 'ideotypes'. What is an ideotype? It is, in effect, the ideal crop plant – the 'perfect model' to guide the selection of superior plus-trees from wild populations. The judges of Miss World competitions probably don't realize it, but they use the same approach to choose the most attractive girl. In the case of developing a new crop, we are of course selecting natural traits that will fulfill the needs of mankind. This is very different from natural selection by Mother Nature, which is selecting the trees that will survive and reproduce under the hazardous conditions of the wild.

Using this ideotype concept, we found that in the bush mango populations there were some individual trees that had big fruits (in length, width and weight) that also had good flesh depth and taste. In the same way there are other trees with large and easily extracted kernels (heavy kernels in nuts with brittle shells). We can therefore say that we have some plants that produce fruits that come close to the ideal fruit ('fruit ideotype') and others that produce kernels that come close to the ideal kernel ('kernel ideotype'). The trees that conform to these ideotypes are therefore potential candidates for propagation as cultivars with very different characteristics – the first step towards crops as different as dog breeds. Likewise the data from safou fruits also identified certain trees with excellent fruit characteristics in terms of size and taste that compared favourably with an ideotype for fresh fruits. Interestingly, it is very rare to find trees that meet both the ideal fruit and kernel criteria.

I will not go into detail here, but we can take this ideotype approach several steps further. For example, the kernel ideotypes could be subdivided to identify candidates for different types of food oils or food thickeners. So, as things progress, the ideotype concept can be used to identify different types of products that are specific to different industries. To ensure that progress towards tree domestication is coordinated and steered in this direction requires dialogue and collaboration between agroforestry researchers and food-industry scientists. The important point to emphasize here is that there is potential for these new tree crops to be highly selected for different market opportunities. In Chapter 9 we will discuss how this can be done responsibly so that the farmers are also beneficiaries from the commercialization process.

I have already mentioned that the kernels of bush mango are used as a thickening agent in traditional soups and stews in West and Central Africa. In our study we wanted to know if this physical characteristic also varied between trees. To assess this, kernels from each tree were stored individually and sent to Campden and Chorleywood Food Research Association in England for chemical and physical analysis. The tests simulated the cooking process and determined the tree-to-tree variation of the 'drawability' of the mucilaginous substance that confers the food-thickening properties of the kernels. Interestingly, the results came back showing that indeed there was 13-fold variation between samples in their drawability and threefold variation in viscosity, a previously unknown food-thickening variable, which was independent of drawability. None of the trees assessed had high values for both viscosity and drawability. This shows that there is great potential to select trees with the best cooking quality, but it raises the question: 'Which is more important to consumers: viscosity or drawability?' The chemical analyses also identified that there was big tree-to-tree variation in oil content of kernels (37.5–75.5%).

The presence of these three traits of likely interest to the food industry indicates the probable need to subdivide the kernel ideotype of bush mango into a food-thickening ideotype based on either viscosity or drawability, or into an oil ideotype. A similar opportunity to develop a hierarchical approach to the ideotype is found in the marula tree, in which both the fruit flesh and the kernels again have a number of potential uses.

Similar opportunities seem to be present in all of the tree species we have examined. In Chapter 8, for example, we will see the same kind of tree-to-tree variation in the medicinal properties of galip nut oil and in the content of four different essential oils in the heartwood of sandalwood. Thus, with regard to tree selection, the ideotype concept helps us to match tree products with different markets, so creating cultivars in any one tree species that are as distinct as dog breeds like a St Bernard, an Afghan hound, a bulldog, a working collie and a toy poodle. The take-home message from all of this is that trees have this inherent variability in all kinds of different selectable traits, so the domestication potential of trees is enormous and, as we have said, virtually untouched by science. New and exciting opportunities for the food industry therefore lie hidden in the Trees of Life. In Chapter 9, we will discuss the appropriate approaches towards commercialization of these products.

One unexpected result from the safou study was that a few trees produced fruits that were seedless. Seedlessness in fruits is often considered a desirable trait in fruits (e.g. grapes, citrus and bananas), and this would probably be the case in safou, as the seed is not used for anything except pig feed. However, there is a commonly held belief in Cameroon that seedless safou are somehow unacceptable or taboo. This of course could pose a problem of market acceptability if this trait was brought into the selection programme without addressing the social issues.

In all food products, taste is obviously a trait of considerable interest, but it is usually difficult to quantify. In Cameroon, Eduard Kengni has made a start by examining variability in acidity, astringency, bitterness and sourness, using a number of consumers as a tasting panel. The results indeed indicate that the fruits from different trees vary in these traits, which should then be added to the ideotype. Likewise, we know in a number of species that there is also tree-to-tree variation in the content of vitamins, minerals and fatty acids, all of which contribute to the nutritional value of fruits. Perhaps this is why domestication is a process that can continue for thousands of years! Currently, we are not getting the farmers involved in this level of sophistication. Nevertheless, malnutrition is one of the key problems that could be addressed by the domestication of indigenous fruit and nut trees, as we know that many of these fruits and nuts are, in contrast to the staple starch foods, rich in nutrients (Table 5.2).

So far in this chapter we have only considered the variability of the fruits and nuts themselves, but of course there is also variation in production traits, such as yield, seasonality and regularity of production, reproductive biology and reduction of susceptibility to pests and diseases.

Many wild fruit trees have variability in their fruiting season, which typically extends over a period of 2–4 months, with individual trees differing in their period of ripening and fruit drop. Within this general pattern, there can be a few trees that fruit outside the normal season, or that fruit more than once per year. Cultivars derived from these trees would extend the productive season. For example, ICRAF has released a cultivar fruiting at Christmas, called 'Noël'. This is of interest as out-of-season fruits can be marketed at high prices. In safou, for example, the price of out-of-season fruits is about ten times the peak-season price. To take advantage of this, a local entrepreneur in Makenene has started to propagate the rare individuals that fruit out of season.

In the preceding section I mentioned that there were some trees of bush mango in Nigeria and some safou trees in Cameroon that had exceptionally large fruits. I wondered if this was evidence of domestication by local communities. In a brainwave one night, I realized that we probably had data that could assess this and I hatched a hypothesis that we could test. The hypothesis was that in totally wild populations the tree-to-tree variability of any measured trait would have a distribution pattern that formed a normal 'bell curve'. But, if farmers were selecting the best trees in their farms, either by cutting out the worst trees, or planting the seeds of their best trees, this would change the shape of the frequency distribution curve. Without going into detail, five distinct stages, each with a specific frequency distribution pattern, would be found as a population progressed from being fully wild to being a domesticated variety. What this means is that we can identify five testable stages in the domestication process that will indicate just how much our measured tree population diverges from the wild state. With my hypothesis tucked under my pillow I went back to sleep.

When the next day I analysed our data in this way, I was excited to find that we did indeed have evidence that in some of our villages the domestication of bush mango and safou had proceeded to Stages 2–3 in Nigeria and Cameroon, respectively. This farmer-driven selection had resulted in a 67% genetic gain in flesh depth in safou and a genetic gain of 44% in flesh depth in bush mango in the most intensively selected village populations. The clinching evidence, however, was that these changes in frequency distribution only occurred in the traits like fruit size and flesh depth, that people would be using to select the best trees, and were completely absent in traits of zero interest, like the kernel size of safou. Since doing this frequency distribution analysis in safou and bush mango, we have also done it in other species such as marula in South Africa and Namibia, and in cutnut in the Solomon Islands, with very similar results.

Before moving on to look at the use of vegetative propagation to capture the variation found in individual trees, I think it is briefly worth mentioning the outcome of some earlier work to assess genetic variation in the timber tree obeche (*Triplochiton scleroxylon*), as it could be relevant to fruit trees. Trees allocate their wood production to the trunk and to branches. In timber trees we are interested in the size of the trunk, but in fruit trees we need the branches on which trees produce fruits. The branching architecture of trees can be categorized into 26 different models, all of which are controlled by a process known as 'apical dominance' – the ability of the shoot tip to regulate the outgrowth of buds on the stem to form branches. Recognizing the importance of this process, we developed a test to assess the tree-to-tree variation in strength of apical dominance in small plants in the nursery. When we later applied this test we found a strong relationship between this simple predictive test and the branching frequency of the same cultivars growing in the field.[2] Potentially, this test, when done under controlled conditions, can save years of expensive field trials by identifying which trees have desired branching characteristics while they are still small plants in the nursery. Some of these same cultivars were also found to differ in their rates of photosynthesis. This characteristic, which determines the production of carbohydrates, is likely to relate to yield potential.

At the start of Chapter 5, I mentioned that tree breeding is a slow process because of the long reproductive cycle. If trees could be induced to flower early, this cycle could be reduced by many years. In obeche, we were successful in developing a technique that induced precocious flowering in young trees just a few months old[3] by making the root system dormant while the shoots were active. This led to carbohydrates accumulating in the shoots and this triggered the induction of flower buds. As these flowers developed we pollinated them with deep-frozen pollen and they produced normal healthy seeds. In this way we achieved three generations of obeche in 7 years, something that would take a minimum of about 45 years in the field. This result indicates that it is possible to overcome the biggest constraint to tree breeding – the long generation time – and so to affect a breeding programme to augment other approaches to tree domestication. However, the current technique, which involves the regulation of root

Fig. 6.3. The stages of cultivar development. (a) The variation of fruits in wild populations; (b) propagation of selected tree by marcotting; (c) use of non-mist propagators to multiply cultivars using stem cuttings; (d) a superior cultivar showing fruits at ground level when the plant is only 18 months old; (e) fruits of the selected cultivar showing their uniformity with the original selected tree.

temperature, is not yet very practical for a large-scale breeding programme, and clearly further work is needed to improve the technique.

Fortunately, tree domestication is not entirely dependent on tree breeding and we have the very powerful option to use asexual reproduction – or vegetative propagation – to create cultivars (Fig. 6.3). This, however, requires a very special set of skills, which are the subject of the next chapter. It is these techniques that allow us to rapidly domesticate trees for their products (timbers, fruits and nuts, etc.). Here we have the crux of how science can take the Trees of Life and turn them into crops that both meet the livelihood needs of local people and the environmental needs of the global community through agroforestry.

Notes

[1] Joseph Kengue, Jean-Marie Fondoun, Zac Tchoundjeu and Paul Anegbeh.
[2] In collaboration with Dr David Ladipo and Professor John Grace.
[3] In collaboration with Dr Alan Longman, Nina Ferguson and Rita Manurung.

Further Reading

Anegbeh, P.O., Usoro, C., Ukafor, V., Tchoundjeu, Z., Leakey, R.R.B., et al. (2003) Domestication of *Irvingia gabonensis*: 3. Phenotypic variation of fruits and kernels in a Nigerian village. *Agroforestry Systems* 58, 213–218.

Anegbeh, P.O., Ukafor, V., Usoro, C., Tchoundjeu, Z., Leakey, R.R.B., et al. (2004) Domestication of *Dacryodes edulis*: 1. Phenotypic variation of fruit traits from 100 trees in southeast Nigeria. *New Forests* 29, 149–160.

Atangana, A.R., Tchoundjeu, Z., Fondoun, J.-M., Asaah, E., Ndoumbe, M., et al. (2001) Domestication of *Irvingia gabonensis*: 1. Phenotypic variation in fruit traits in 52 trees from two populations in the humid lowlands of Cameroon. *Agroforestry Systems* 53, 55–64.

Atangana, A.R., Ukafor, V., Anegbeh, P., Asaah, E., Tchoundjeu, Z., et al. (2001) Domestication of *Irvingia gabonensis*: 2. The selection of multiple traits for potential cultivars from Cameroon and Nigeria. *Agroforestry Systems* 55, 221–229.

Barany, M., Hammett, A.L., Leakey, R.R.B. and Moore, K.M. (2003) Income generating opportunities for smallholders affected by HIV/AIDS: linking agro-ecological change and non-timber forest product markets. *Journal of Management Studies* 39, 26–39.

Jamnadass, R.H., Dawson, I.K., Franzel, S., Leakey, R.R.B., Mithöfer, D., et al. (2011) Improving livelihoods and nutrition in sub-Saharan Africa through the promotion of indigenous and exotic fruit production in smallholders' agroforestry systems: a review. *International Forestry Review* 13, 338–354.

Leakey, R.R.B. (2005) Domestication potential of marula (*Sclerocarya birrea* subsp. *caffra*) in South Africa and Namibia: 3. Multi-trait selection. *Agroforestry Systems* 64, 51–59.

Leakey, R.R.B. (2007) Domestication and marketing of novel crops. In: Scherr, S.J. and McNeely, J.A. (eds) *Farming with Nature: the Science and Practice of Ecoagriculture.* Island Press, Washington, DC, pp. 83–102.

Leakey, R.R.B. and Page, T. (2006a) Domestication of agroforestry trees: progress towards adoption. *Forests, Trees and Livelihoods* 16, 1–121.

Leakey, R.R.B. and Page, T. (2006b) 4 + 4 + 4 = progress. *Forests, Trees and Livelihoods* 16, 3–4.

Leakey, R.R.B. and Page, T. (2006c) The 'ideotype concept' and its application to the selection of 'AFTP' cultivars. *Forests, Trees and Livelihoods* 16, 5–16.

Leakey, R.R.B., Fondoun, J.-M., Atangana, A. and Tchoundjeu, Z. (2000) Quantitative descriptors of variation in the fruits and seeds of *Irvingia gabonensis. Agroforestry Systems* 50, 47–58.

Leakey, R.R.B., Atangana, A.R., Kengni, E., Waruhiu, A.N., Usuro, C., *et al.* (2002) Domestication of *Dacryodes edulis* in West and Central Africa: characterisation of genetic variation. *Forests, Trees and Livelihoods* 12, 57–71.

Leakey, R.R.B., Tchoundjeu, Z, Smith, R.I., Munro, RC., Fondoun, J-M., *et al.* (2004) Evidence that subsistence farmers have domesticated indigenous fruits (*Dacryodes edulis* and *Irvingia gabonensis*) in Cameroon and Nigeria. *Agroforestry Systems* 60, 101–111.

Leakey, R.R.B., Tchoundjeu, Z., Schreckenberg, K., Shackleton, S. and Shackleton, C. (2005a) Agroforestry tree products (AFTPs): targeting poverty reduction and enhanced livelihoods. *International Journal of Agricultural Sustainability* 3, 1–23.

Leakey, R.R.B., Shackleton, S. and du Plessis, P. (2005b) Domestication potential of marula (*Sclerocarya birrea* subsp. *caffra*) in South Africa and Namibia: 1. Phenotypic variation in fruit traits. *Agroforestry Systems* 64, 25–35.

Leakey, R.R.B., Pate, K. and Lombard, C. (2005c) Domestication potential of marula (*Sclerocarya birrea* subsp. *caffra*) in South Africa and Namibia: 2. Phenotypic variation in nut and kernel traits. *Agroforestry Systems* 64, 37–49.

Leakey, R.R.B., Greenwell, P., Hall, M.N., Atangana, A.R., Usoro, C., *et al.* (2005d) Domestication of *Irvingia gabonensis*: 4. Tree-to-tree variation in food-thickening properties and in fat and protein contents of dika nut. *Food Chemistry* 90, 365–378.

Leakey, R.R.B., Tchoundjeu, Z., Schreckenberg, K., Simons, A.J., Shackleton, S., *et al.* (2007) Trees and markets for agroforestry tree products: targeting poverty reduction and enhanced livelihoods. In: Garrity, D., Okono, A., Grayson, M. and Parrott, S. (eds) *World Agroforestry into the Future.* World Agroforestry Centre, Nairobi, pp. 11–22.

Leakey, R.R.B., Weber, J.C., Page, T., Cornelius, J.P., Akinnifesi, F.K., *et al.* (in press) Tree domestication in agroforestry: progress in the second decade (2003–2012). In: Nair, P.K. and Garrity, D. (eds) *Agroforestry: The Way Forward.* Springer Verlag, Dordrecht, The Netherlands.

Tchoundjeu, Z., Kengue, J. and Leakey, R.R.B. (2002) Domestication of *Dacryodes edulis*: state-of-the art. *Forests, Trees and Livelihoods* 12, 3–13.

Waruhiu, A.N., Kengue, J., Atangana, A.R., Tchoundjeu, Z. and Leakey, R.R.B. (2004) Domestication of *Dacryodes edulis*: 2. Phenotypic variation of fruit traits in 200 trees from four populations in the humid lowlands of Cameroon. *Food, Agriculture and Environment* 2, 340–346.

Vegetative Propagation 7

> The propagation of plants is a fundamental occupation of mankind.
> Hudson Hartmann and Dale Kester (1968) *Plant Propagation*.
> Prentice-Hall, New Jersey.

> Recent systematic approaches to optimization of pre and post severance environmental conditions for the rooting of cuttings have demonstrated that cutting propagation is feasible for a much broader range of multipurpose and fruit tree species than previously considered.
> Kenneth Mudge and Eric Brennan (1999) Clonal propagation of multipurpose and fruit trees used in agroforestry. In: Louise Buck *et al.* (eds) *Agroforestry in Sustainable Agricultural Systems*. CRC Press and Lewis Publishers, New York.

Vegetative propagation is the most effective way to rapidly capture and harness the superior traits of genetically superior plants in wild populations. The techniques involve the creation and multiplication of new plants without sexual reproduction, for example by rooting cuttings, grafting or marcotting. These techniques create a genetically identical and exact copy of the plant from which the cuttings or buds are taken. This creates what is known as a clone. If the clone originates from a tree selected for its superiority in yield, quality or other desirable trait – what we call a 'plus-tree' – then that clone becomes a potential crop variety or 'cultivar' for further field testing and selection during the domestication process. So vegetative propagation is a powerful tool for domestication and we need to understand how to use these techniques, as without them the domestication of long-lived trees is a very slow process dependent on a breeding programme.

We are all much more familiar with clones than we may actually realize, as we see and handle their products every day whenever we buy a banana, a potato, a rose or an apple. All these domesticated species have been propagated vegetatively.

Before we go on, perhaps I need to give some explanation here of how vegetatively propagated trees differ from trees grown from seeds. Seeds are the outcome of sexual reproduction, which involves the fusing together of different bits of the genetic material of both parents. That is why seedlings, like children, are not copies of either parent, nor are they copies of their siblings – brothers and sisters.

Unlike domestication by vegetative propagation, domestication by tree breeding is a slow process because each generation has to grow for many years and mature before it can in turn produce progeny, and not all the individuals in this progeny will display the characteristics of the parents. So the average performance of the progeny may only be marginally better than those of the parents.

As I mentioned earlier, if you want to see how individual tree clones differ from each other, you can plant out lines of different clones in a field trial. As they grow, first you will very clearly see the similarities within the plants of the clone and the differences among the different clones in terms of their height, branching pattern, and leaf, bark or flower colour, etc. Secondly, by making measurements of how these things relate to the yield, or the shape of the trunk, you can determine which characteristics are useful predictors of the growth and development of the tree, and so how they are likely to influence the performance and yield of the clone. This of course is the sort of information that foresters, agroforesters and horticulturalists want to know when wishing to domesticate and develop the species as a crop.

Twenty to thirty years ago vegetative propagation of tropical trees was a skill largely attributable to 'green fingers', as basic principles applicable to all trees were not well understood and many species were thought to be impossible to propagate. This was a serious problem, as vegetative propagation is needed to create tens, hundreds, thousands or even millions of copies of an individual plant.

In this chapter I am going to introduce some principles derived from our research.[1] They are based on simple and robust techniques that can be used to create a cultivar, even in remote corners of tropical countries with minimal infrastructure. We will start with the techniques of grafting, budding and marcotting, which are especially useful to capture the unique combination of traits found in a selected and sexually mature 'plus-tree'. In fruit and nut trees, the capture of the mature state is important, as we want our selected cultivar to have the capacity to flower and form fruits at an early age.

So, how does a tree become mature? As a general principle, mature trees are large and at least 10–20 years old. But there the similarity between trees and people stops. In animals, like people, the whole body of an individual changes when it becomes sexually mature. In trees, however, the juvenile seedling slowly matures as it gets bigger and bigger until it reaches a threshold when the newly formed shoots are mature and have the capacity to flower and fruit. The bottom and older shoots of the tree, which were formed before reaching this threshold, remain juvenile (immature). This means that to propagate mature trees we have to take the tissues we need from the mature crown. These tissues are very difficult to propagate by cuttings, as we will see later, but are more amenable to grafting, budding and marcotting.

The propagation of mature tissues overcomes one of the big constraints always mentioned by farmers when you discuss tree crops – they want quick returns from a new crop and do not want to wait many years while it grows and matures. To achieve this we have to propagate mature shoots. Then, once the mature tree material has been brought into the nursery as a grafted plant or a rooted marcott, it can be rapidly and easily multiplied by rooting leafy stem cuttings.

The existence of mature and juvenile tissues in the same tree actually gives us a unique opportunity to manipulate the size and time to first fruiting in cultivars. Tall, immature trees with a long trunk result from the propagation of juvenile shoots, while short, mature trees without a trunk result from propagation of mature shoots (Fig. 7.1). In effect, by choosing whereabouts on the tree the propagation material is taken, you can 'design' the stature and precocity of the tree you are going to produce. So, shoots from the top of the crown will result in small fruiting trees for planting in a homestead garden or orchard; shoots from just below the threshold to maturation will result in taller trees suitable for certain crop mixtures; while juvenile material from the base of the trunk will result in a full-sized tree to provide shade and timber.

Grafting techniques are relatively simple and have been around for hundreds, if not thousands of years. The technique involves detaching a

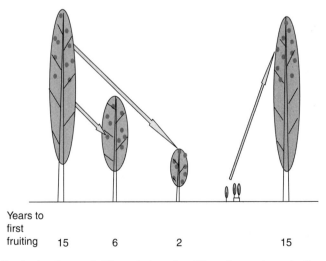

Years to first fruiting   15   6   2   15

Fig. 7.1. Designing trees of different sizes for different uses by selecting the source of the propagation material vis-à-vis the state of maturation in the tree (see text for full explanation).

shoot (scion) from a selected desirable tree and carefully reattaching it to an unselected seedling so that the different internal tissues of the scion and the rootstock are juxtaposed and heal over, allowing the scion to grow and form a new tree. Creating new plants in this way requires some skill. In addition, care has to be taken to protect the graft union from drying out or from any movement. Furthermore, the rootstock can reject the scion. This incompatibility problem can arise when unrelated trees are grafted. It is rather like the 'tissue rejection' we see in a kidney/lung/heart transplant, and it typically results in the death of the cultivar. This rejection can take 10 or more years to occur and so can be a serious setback to a domestication programme.

Marcotting (Fig. 7.2) – or air layering – is somewhat easier, but the multiplication rates are generally lower. The technique involves the development of roots on a shoot within the tree crown (preferably a vertical shoot and not a horizontal branch), while that shoot is still part of

Fig. 7.2. A marcott set on a shoot to form roots.

the mother tree. To stimulate rooting on an intact shoot you cut out a ring of bark (10–15 mm wide) close to where two shoots join, apply rooting hormone and then wrap up the wound with black plastic, enclosing a plug of damp compost, and tie it tight top and bottom. Then once it is rooted you sever the shoot and pot it up in the nursery. This can be a very effective way to capture the mature attributes of a superior tree and create a plant that can subsequently be grown in the nursery as a stockplant for mass propagation by stem cuttings.

The rooting of cuttings is the easiest and most effective way to mass produce plants vegetatively, especially if the source of the cuttings is juvenile trees (i.e. from seedlings, saplings and coppiced stumps, hedges or regularly pruned stockplants). It is this ability to propagate trees vegetatively that is the key that has now unlocked our ability to enrich agroforestry systems with high quality cultivars of indigenous fruit and nut trees, and so open up opportunities for generating income from village-level tree nurseries.

Thanks to my earlier work with tropical hardwoods,[2] especially 'obeche', we now have a good understanding of some basic principles that make it relatively easy to root stem cuttings of the species that farmers are nominating for domestication. The result of this is that vegetative propagation techniques have, I believe, now been converted from a green-fingered art to a science. Our research was all based on single-node stem cuttings (Fig. 1.5 – this is a rooted cutting) – a piece of a shoot from a seedling or regularly pruned stockplant with a single bud and a trimmed leaf.

Because of the importance of vegetative propagation as a tool in tree domestication I am going to describe five key factors determining success. Some of these factors are associated with the leaf and determine the rates of producing sugars by photosynthesis, or the rate of water loss by transpiration. Other factors are associated with the stem and determine the quantity and availability of stored nutrients, carbohydrate reserves or water, as well as the woodiness of the stem. Together these leaf and stem factors affect the capacity of the cutting base to produce and develop new root tissues.

As we will see, to achieve success in vegetative propagation requires the avoidance of stress and the maximization of physiological vigour – a bit like preparing a squad for a major sporting event like the Olympics! As soon as a cutting is severed from the stockplant, it is without roots and so without a source of water to keep its cells and tissues fully hydrated. If water stress occurs, the leaf will wilt and probably drop off. Leaves respond to the loss of water (water stress) by closing their numerous tiny stomata, which we can think of as breathing holes. By closing the stomata the leaf conserves water, but this occurs at the expense of shutting down photosynthesis, the process by which plants convert carbon dioxide and water into sugars. So if the stomata are closed for a long time, the quantity of carbohydrates in the leaf and stem will decline as they are burnt up in the process of respiration: the process that releases energy for all the functions of life. These are the basic physiological functions of plant growth and if we are to induce our cutting to root, we have to do everything we can to keep these processes active so that cells at the base of the stem can divide and organize themselves into root tissues, which then elongate and emerge as new roots. In essence, therefore, successful propagation involves maintaining a stress-free environment throughout the period of propagation. I'm sure the same thing is true for athletes before and during the Olympics.

Critical Factor 1: the Propagation Environment

Success depends on keeping the detached shoots moist and cool from the moment they are severed from the stockplant. Each shoot is then quickly cut up into single-node pieces, given any treatments needed (see Critical Factor 2) before being firmed into the rooting medium (typically sand, fine gravel, peat, perlite, vermiculite or various mixtures of them) inside a propagator. There are many different ways of providing an appropriate aerial environment for stem cuttings that vary in cost and sophistication,[3] but the most common are automated systems that generate a fine spray of mist or a fog of tiny water drops. These of course are dependent on an electricity supply. To circumvent this requirement in remote tropical locations, we developed simple non-mist propagators that did not require electricity or piped water, and were also cheap to make and simple to operate (Fig. 7.3). Subsequent testing showed that these propagators were equally efficient, if not better than the mist system, and had much lower maintenance requirements. Their simplicity has made them very popular, and they can now be found in use all around the world.

Basically, the non-mist propagator is an airtight, watertight box made from polythene (good quality plastic sheeting) stretched around a wooden frame (Fig. 7.4)[4] – well actually it can be made of anything, and I have seen modifications made of bamboo and metal framework, coming in all sorts of shapes and sizes. The exact construction is therefore unimportant, what matters is that the propagator can hold a body of water without leaking, and that the air above the rooting medium has a relative humidity close to 100%. This is achieved by filling the base of the watertight frame with stones, topped with gravel, to a depth of about 15–20 cm and then filling the box with water up to the surface of the gravel. Then a 10 cm layer of rooting medium is placed on top of the gravel (Fig. 7.3). The rooting medium absorbs some moisture from the stone and gravel layer, which keeps it moist without saturating it. There is also some evaporation from the surface of the rooting medium, which humidifies the air in the top of the propagator frame.

Fig. 7.3. Cross-section of a non-mist propagator showing the different layers, the water table (arrowed) and rooted cuttings in the rooting medium.

Once the cuttings have been inserted into the propagator, the lid is closed with an airtight seal so that the humidity builds up rapidly. The lid should then be kept closed. This provides a near perfect stress-free environment for cuttings so the stomata can reopen and the physiological processes can occur normally. Inside this simple structure, the severed cuttings can produce new carbohydrates and exchange carbon dioxide, oxygen and water normally and they've got a very good chance of developing a set of vigorous new roots.

Regardless of the type of propagator, the composition of the rooting medium is often critical for rooting. In addition to holding the cutting firm, it has to provide moisture and air. A lack of air at the cutting base encourages rotting, which is one of the common causes of failure or low rooting success. In cold temperate environments propagation beds typically have some 'bottom-heat' provided by heating pipes or cables to promote cellular activity at the cutting base, while the leaves remain cool.

Critical Factor 2: Post-severance Treatments

There are a number of ways that cuttings can be treated to maximize rooting success. The most widely recognized is the use of rooting 'hormones' (auxins) to enhance both the speed of rooting and the number of roots

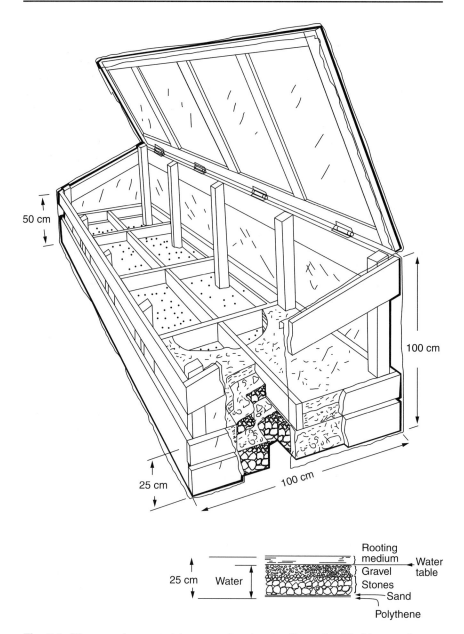

Fig. 7.4. Diagram of a non-mist propagator showing the water-filled layers of stones and gravel, which are topped with the rooting medium. (Source: Longman, 1993.)

formed. In our human analogy of sporting prowess, perhaps auxins are like performance-enhancing drugs. Auxins are applied to the base of cuttings to stimulate the developmental processes of cell division and the formation of new root tissues. As a general rule, a droplet (10 µl) of a 0.4% solution dissolved in alcohol is optimal. Auxins seem to act like a magnet attracting nutrients and carbohydrates to the cutting base where the new roots will form. Other post-severance treatments that can be manipulated to test their effects on rooting are leaf area and stem length. In sporting terms, perhaps this equates to what the physical characteristics of good athletes are and how they can be prepared to give their best performance. Auxins are, however, interactive with other variables like the environment and the quality of the cutting material (see Critical Factors 4 and 5). Failure to recognize this probably accounts for some species being reported to be non-responders to auxin. This equates with the failure of performance-enhancing drugs to turn an unfit weakling into an Olympic athlete.

Typically, the rooting of softwood cuttings is dependent on the presence of a leaf to provide the necessary sugars. We found, for example, that a leafless cutting will use up its stored carbohydrate reserves in about 9 days, while a leafy cutting, with the optimal leaf area, will double its carbohydrate reserves in the same period. The larger the area of leaf, the more carbohydrates the leaf can produce, but large leaves have a cost, in terms of their need for water. This means that there is a trade-off and the optimal size of the leaf is that which balances the positive effects of photosynthesis producing sugars to support cell growth, with the negative effects of water loss creating water stress. As a general rule, a leaf area around 50cm^2 is optimal. This can be achieved by trimming the cutting leaf with scissors. The other important aspect of this is that the cutting has the ability to produce carbohydrates from day one. If the cutting is stressed, it closes its stomata to reduce water loss and so cannot produce sugars by photosynthesis.

As a general rule, long cuttings root better than short ones; we are talking about cuttings in the range of 0.5–15.0 cm. However, we usually use single-node cuttings (Fig. 1.5) because this maximizes the number of plants that can be produced. Their length and diameter is affected by their position within a shoot and within the stockplant (Critical Factor 3) and also the distance between leaves on a stem (the internode length). The latter is attributed to the effects of the stockplant environment on the rate of shoot growth at the time that each leaf was formed. Consequently the rooting ability of cuttings also varies depending on the original position of the cutting within the stockplant. This means that stockplant management is important.

Critical Factor 3: Stockplant Factors Within a Single Shoot

In addition to the above mentioned effects of position on internode length, there are age-related gradients from top to bottom in leaf size, bark colour and the woodiness of the stem. Measurements show that there are also differences in carbohydrate, nutrient and water content. This means that no two cuttings from a single plant are the same physically or nutritionally, although they are genetically identical. Our research has found that it is the variation in the internode length which is the most important of these age-related variables in terms of their effects on rooting capacity. So, as in an athlete, the physique is important. Further work found that small cuttings lack sufficient storage space for newly formed sugars. With nowhere for the sugars to go, the cutting switches off its photosynthetic processes. It's the cutting's equivalent of our athlete having eaten too much Christmas dinner!

Critical Factor 4: Stockplant Factors Between Shoots

Now we are getting into fairly uncharted territory. To try to provide some understanding we initiated a series of innovative experiments which did generate new information. To add to what we saw in Critical Factor 3, we investigated what happens when a stockplant has more than one shoot and found that cuttings taken from different shoots on even simple stockplants with two to four shoots rooted very differently – between 0 and 100%, but with some very clear patterns of variation. For example, stockplants that varied in height from 30 to 100 cm produced between two and seven shoots, respectively. When cuttings were taken from the top shoot in each case, rooting success was low in cuttings from the tallest plants (20%) and high in cuttings from the shortest plants (84%). A possible explanation of this is that the shoots on a stockplant compete for something they need – perhaps for water or nutrients. Certainly when I was at school, if you wanted to get more food, you had to sit near to the person serving the food and compete with the other boys wanting seconds. When we tested this hypothesis by controlling the number of shoots on plants of the same height we found some evidence supporting the idea of competition but it was not conclusive.

Digging a bit deeper, we found that lower shoots were not affected by competition from above, while upper shoots were affected by competition from below. So what we see here is that when stockplants had the same number of shoots, there was a positional effect favouring those from the bottom of the stockplant (68–72%), rather than those from the top

(24–42%). An explanation for this became apparent when we repeated this experiment, but arranged for the lower and upper shoots to be in the same light environment. In this case, lo and behold, the positional effect went away (66% from both positions). These results therefore suggest that the positional effect is actually more a result of the light environment experienced by the shoot than its distance from the roots. These results were very surprising – even exciting! This was the first time that this role of light had been demonstrated. It was particularly exciting as at the time we did this work we had no reason to expect that light was such a powerful factor affecting the rooting of cuttings. So what is the role of light in all this?

Critical Factor 5: the Stockplant Environment

Before progressing, we need to understand light and its role in plant growth and development. Light is of course important for photosynthesis, with the rates of sugar production being affected by the amount of available light. The amount of light is regulated by the power of the light source, whether it is the sun or artificial lighting; as well as by interception, which causes shading. However, in addition to varying in quantity, light can vary in quality. Plants are very responsive to light quality, as it has an important role in plant development. This means that in response to the quality of light reaching their leaves, plants control the way they grow by regulating the allocation of carbohydrates to different tissues (stem, leaf or root), so affecting their subsequent growth. This development process determines whether the piece of stem between two leaves and buds (the internode) is long or short, whether a leaf is small or large, and whether a leaf is thin and soft or thicker and more rigid. To explain this I need to introduce 'phytochrome'. It is a pigment in plant cells that is sensitive to specific wavelengths of light – called 'red' and 'far-red'. Consequently, when a plant receives light this pigment assesses the amount of red and far-red light (the ratio of red to far-red) being received and accordingly regulates the allocation of carbohydrates and other products to the different tissues.

In nature, the red to far-red ratio of light is modified as light passes through a green leaf, as the red light is filtered out more than other wavelengths. The more leaves the light passes through the greater is the amount of red light removed. Thus when parts of a plant are in the shade of either their own leaves or the leaves of neighbouring plants two things happen: (i) the amount of light is reduced; and (ii) the quality of the light is changed (the red:far-red ratio is lower).

To investigate the role of light that we saw in the previous experiment, we wanted to test independently the effects of both the amount of light (irradiance) and the quality of the light (red:far-red ratio) on the performance of our cuttings. Technically it is quite difficult to vary the different levels of irradiance without changing the light quality, and conversely to vary light quality without changing irradiance. Nevertheless, we did manage to do these two experiments using controlled environment growth chambers, and the experiments clearly showed for the first time that growing stockplants under different levels of irradiance and different light qualities both independently affected the rooting success of cuttings. In both instances, the treatments closest to shade light were associated with the best rooting, thus in nature these two independent characteristics of shade light work additively to greatly enhance the rooting ability of cuttings taken from shaded shoots. This therefore supports the findings in Critical Factor 4 that the effects of shoot position on rooting could be due to their different light environments.

How does this come about? Basically, we found that the cuttings grown under shade, in terms of both light quality and irradiance, had longer stems and bigger leaves, as well as being generally taller and less 'bushy' plants. Together with the earlier findings, this was a breakthrough in terms of understanding how to manipulate stockplants and improve the success of vegetative propagation.

Before going on with this story, I am going to introduce another complicating factor. It is the role of plant nutrients. We knew that cuttings taken from stockplants grown with artificial fertilizers rooted better on average than cuttings taken from stockplants without fertilizers. Not a very surprising result. However, there was a big surprise hidden within those results. When we analysed the data shoot by shoot we found that the benefit of fertilizers was limited to the shaded bottom shoots and had no effect on the vigorous upper shoots. Later we showed that these beneficial shade + nutrients effects were attributable to a preconditioning effect in the stockplant, which boosted rates of photosynthesis and lowered the starch content of the shoots. These results were later confirmed in other species. In our Olympics analogy, this preconditioning effect is like providing an athlete with a healthy home environment and good nutrition.

This interaction between shade and nutrients has very important practical implications for stockplant management as a tool to precondition shoots for enhanced rooting ability which we have used many times. We have found, for example, that growing stockplants in the shade of nitrogen-fixing trees is a very effective way to precondition the shoots so that they root easily when taken as cuttings.

I have made several comparisons between successful propagation and successful performance in a sporting event like the Olympics. To summarize this, both stem cuttings and athletes have to: (i) be physiologically fit;

(ii) have the ideal physique; and (iii) be stress-free, lean, well prepared and energetic. They also have to be highly competitive and ready for action.

Propagation of Mature Trees

At the start of this chapter, I mentioned that the propagation of mature trees is much more difficult than propagation from seedlings, coppice shoots or managed stockplants. It would be highly desirable to be able to propagate mature trees by taking cuttings. This raises the question as to whether the problem is due to the maturation process or something else. Looking at Critical Factors 3, 4 and 5, it seems very clear that cuttings from the crown of a large tree combine all the undesirable characteristics. In a study with bird cherry (*Prunus avium*)[5] our results confirmed that all the differences in the size of the stems and leaves were big enough to explain the poor rooting of the cuttings from the tree crown. So the differences in rooting responses between 'juvenile' and 'mature' were indeed predictable, without having to evoke a 'maturity' factor to explain the differences.

To provide more definitive evidence we need to somehow produce shoots up in the crown of a big tree that had the form and characteristics of 'juvenile' cuttings; and then compare their rooting responses with both normal mature crown shoots from the same part of the tree crown. I will come back to this in the next chapter.

Genetic Variation in Rooting

It is frequently argued that genetic variation in rooting ability can occur both at the level of individual species and between different clones of a given species. Once again, our evidence suggests that this variation can in fact be explained by small differences in the size of leaves, internodes, etc. (i.e. all the things we saw to affect rooting in Critical Factors 2, 3, 4 and 5), rather than by some gene specifically affecting the ability to form roots. The practical implication of this finding is that more attention needs to be paid to stockplant management to overcome these apparent genetic differences. If species differ in ways that result in them requiring slight modifications to their preparation for propagation, then in our Olympics analogy, species are something like the participants in the different sports that make up the overall competition.

Principles Determining Rooting Success

The scientific literature contains many contradictory statements about how to root cuttings. These arise, I believe, because many researchers do not present information about how they have addressed the five Critical Factors. I hope the information presented in this chapter provides some principles that can be used to guide work with new species.

Another issue is that the results of vegetative propagation experiments are presented as the 'percentage of cuttings rooted' as a measure of success. The problem here is that it is a very poor way of measuring success, because as we have seen in this chapter success is affected by which shoots cuttings are taken from and how the stockplant has been managed. Consequently, the percentage of cuttings rooting is only meaningful if you know which shoots were used and the environmental conditions prevailing at the time, and few researchers give this information.

To illustrate this problem, let's consider a situation in which the propagation environment (Critical Factor 1) and the post-severance treatments (Critical Factor 2) are all optimized. Now, if the person taking the cuttings takes as many of them as possible from a small managed stockplant regardless of their position within or between shoots (Critical Factors 3 and 4), they are likely to have a fairly low rooting percentage (52%) resulting in 16 plants. This is because of the inclusion of all of the inherent variation in cutting size, shoot position, etc. However, if on the other hand, the same person goes to an identical stockplant knowing that only cuttings from the top two shoots are likely to have the capacity to root, and takes just 16 cuttings from these two shoots, he/she will root them all, so now the rooting percentage is 100%. This is then further complicated if, in a third example, the person doing the propagation has a good understanding of the importance of stockplant management, and grows the stockplants under shade light with adequate fertilizers. In this case there may be more shoots from which to take cuttings and these cuttings will be preconditioned to root well. So if 45 cuttings can be taken, including some from a third shoot, and 40 of them root we end up with more plants but the rooting percentage is lower (89%).

To conclude this chapter, I want to reiterate that by overcoming the constraint to tree domestication that was imposed by the difficulties of vegetative propagation we now have principles that, I believe, have converted these vegetative propagation techniques from a green-fingered art to a science. So we now have the key to unlock the domestication process and hence to enrich agroforestry systems with new crops developed from the Trees of Life. Further details are now available as narrated PowerPoint slides as part of a tree domestication training course on the ICRAF website.[6] The serious constraint now to greatly scaling up the domestication of the Trees of Life is the number of people with the skills of vegetative propagation.

Notes

[1] Over the years the team included: Drs Alan Longman, Adrian Newton, Jan Dick, Mike Coutts and Steve Hoad. It also included the following PhD and MSc students:

Francisco Mesen, Zac Tchoundjeu, Aminah Hamzah, Patrick Shiembo, Richard Pauku, Hassan Mohammed, Daniel Ofori and Theresa Nketiah.
[2] Supported by Vicky Chapman, Sabina Knees, Nina Ferguson, Debbie Denovan, Ken East, Colin McBeath, Richard Storeton-West, Ray Ottley and Frank Harvey.
[3] Hartmann, H.T., Kester, D.E., Davis, F.T. and Geneve, R.L. (1997) *Plant Propagation: Principles and Practices*, 6th edn. Prentice-Hall, Upper Saddle River, New Jersey, 770 pp.
[4] Longman, K.A. (1993) *Rooting Cuttings of Tropical Trees*. Tropical Trees: Propagation and Planting Manuals, Vol. 1. Commonwealth Science Council, London, p. 64.
[5] In collaboration with Dr Jan Dick.
[6] www.worldagroforestry.org/Units/training/downloads/tree_domestication

Further Reading

Aminah, H., Dick, J.McP., Leakey, R.R.B., Grace, J. and Smith, R.I. (1995) Effect of indole butyric acid (IBA) on stem cuttings of *Shorea leprosula*. *Forest Ecology and Management* 72, 199–206.

Dick, J.McP. and Leakey, R.R.B. (2006) Differentiation of the dynamic variables affecting rooting ability in juvenile and mature cuttings of cherry (*Prunus avium*). *Journal of Horticultural Science and Biotechnology* 81, 296–302.

Dick, J.McP., Leakey, R., McBeath, C., Harvey, F., Smith, I.R., *et al.* (2004) Influence of nutrient application rate on the growth and rooting potential of the West African hardwood *Triplochiton scleroxylon*. *Tree Physiology* 24, 35–44.

Hoad, S.P. and Leakey, R.R.B. (1994) Effects of light quality on gas exchange and dry matter partitioning in *Eucalyptus grandis* W. Hill ex Maiden. *Forest Ecology and Management* 70, 265–273.

Hoad, S.P. and Leakey, R.R.B. (1996) Effects of pre-severance light quality on the vegetative propagation of *Eucalyptus grandis*. Cutting morphology, gas exchange and carbohydrate status during rooting. *Trees* 10, 317–324.

Leakey, R.R.B. (1983) Stockplant factors affecting root initiation in cuttings of *Triplochiton scleroxylon* K. Schum., an indigenous hardwood of West Africa. *Journal of Horticultural Science* 58, 277–290.

Leakey, R.R.B. (1985) The capacity for vegetative propagation in trees. In: Cannell, M.G.R. and Jackson, J.E. (eds) *Attributes of Trees as Crop Plants*. Institute of Terrestrial Ecology, Abbots Ripton, Huntingdon, UK, pp. 110–133.

Leakey, R.R.B. (1990) *Nauclea diderrichii*: rooting of stem cuttings, clonal variation in shoot dominance and branch plagiotropism. *Trees* 4, 164–169.

Leakey, R.R.B. (1992) Enhancement of rooting ability in *Triplochiton scleroxylon* by injecting stockplants with auxins and a cytokinin. *Forest Ecology and Management* 54, 305–313.

Leakey, R.R.B. (2004) Physiology of vegetative reproduction. In: Burley, J., Evans, J. and Youngquist, J.A. (eds) *Encyclopaedia of Forest Sciences*. Academic Press, London, pp. 1655–1668.

Leakey, R.R.B. and Coutts, M.P. (1989) The dynamics of rooting in *Triplochiton scleroxylon* K. Schum. cuttings: their relation to leaf area, node position, dry weight accumulation, leaf water potential and carbohydrate composition. *Tree Physiology* 5, 135–146.

Leakey, R.R.B. and Mohammed, H.R.S. (1985) Effects of stem length on root initiation in sequential single-node cuttings of *Triplochiton scleroxylon* K. Schum. *Journal of Horticultural Science* 60, 431–437.

Leakey, R.R.B. and Simons, A.J. (2000) When does vegetative propagation provide a viable alternative to propagation by seed in forestry and agroforestry in the tropics and sub-tropics? In: Wolf, H. and Arbrecht, J. (eds) *Problem of Forestry in Tropical and Sub-tropical Countries – the Procurement of Forestry Seed – the Example of Kenya*. Ulmer Verlag, Germany, pp. 67–81.

Leakey, R.R.B. and Storeton-West, R. (1992) The rooting ability of *Triplochiton scleroxylon* K. Schum. cuttings: the interactions between stockplant irradiance, light quality and nutrients. *Forest Ecology and Management* 49, 133–150.

Leakey, R.R.B., Chapman, V.R. and Longman, K.A. (1982) Physiological studies for tropical tree improvement and conservation. Some factors affecting root initiation in cuttings of *Triplochiton scleroxylon* K. Schum., an indigenous hardwood of West Africa. *Forest Ecology and Management* 4, 53–66.

Leakey, R.R.B., Mesén, J.F., Tchoundjeu, Z., Longman, K.A., Dick, J.McP., *et al.* (1990) Low-technology techniques for the vegetative propagation of tropical trees. *Commonwealth Forestry Review* 69, 247–257.

Leakey, R.R.B., Newton, A.C. and Dick, J.McP. (1994) Capture of genetic variation by vegetative propagation: processes determining success. In: Leakey, R.R.B. and Newton, A.C. (eds) *Tropical Trees: the Potential for Domestication and the Rebuilding Forest Resources*. HMSO, London, pp. 72–83.

Longman, K.A. (1993) *Rooting Cuttings of Tropical Trees*. Tropical Trees: Propagation and Planting Manuals, Vol. 1. Commonwealth Science Council, London, 132 pp.

Mathias, P.J., Alderson, P.G. and Leakey, R.R.B. (1989) Bacterial contamination in cultures of tropical hardwood species. *Acta Horticulturae* 212, 43–48.

Mesén, F.J., Newton, A.C. and Leakey, R.R.B. (1997) The effects of propagation environment and foliar area on the rooting physiology of *Cordia alliodora* (Ruiz and Pavon) Oken cuttings. *Trees* 11, 404–411.

Mesén, F.J., Newton, A.C., Leakey, R.R.B. and Grace, J. (1997) Vegetative propagation of *Cordia alliodora* (Ruiz and Pavon) Oken: the effects of IBA concentration, propagation medium and cutting origin. *Forest Ecology and Management* 92, 45–54.

Mesén, F.J., Leakey, R.R.B. and Newton, A.C. (2001) The influence of stockplant environment on morphology, physiology and rooting of leafy stem cuttings of *Albizia guachapele*. *New Forests* 22, 213–227.

Newton, A.C., Dick, J.McP., McBeath, C. and Leakey, R.R.B. (1996) The influence of R:FR ratio on the growth, photosynthesis and rooting ability of *Terminalia spinosa* Engl. and *Triplochiton scleroxylon* K. Schum. *Annals of Applied Biology* 128, 541–556.

Nketiah, T., Newton, A.C. and Leakey, R.R.B. (1998) Vegetative propagation of *Triplochiton scleroxylon* K. Schum in Ghana. *Forest Ecology and Management* 105, 99–105.

Nketiah, T., Newton, A.C. and Leakey, R.R.B. (1999) Vegetative propagation of *Triplochiton scleroxylon* in Ghana: effects of cutting origin. *Journal of Tropical Forest Science* 11, 512–515.

Ofori, D.A., Newton, A.C., Leakey, R.R.B. and Cobbinah, J.R. (1996) Vegetative propagation of *Milicia excelsa* Welw. by root cuttings. *Journal of Tropical Forest Science* 9, 124–127.

Ofori, D.A., Newton, A.C., Leakey, R.R.B. and Grace, J. (1996) Vegetative propagation of *Milicia excelsa* Welw. by leafy stem cuttings. II. Effects of auxin concentration, leaf area and rooting medium. *Forest Ecology and Management* 84, 39–48.

Ofori, D.A., Newton, A.C., Leakey, R.R.B. and Grace, J. (1997) Vegetative propagation of *Milicia excelsa* Welw. by leafy stem cuttings. I. Effects of maturation, coppicing, cutting length and position on rooting ability. *Journal of Tropical Forest Science* 10, 115–129.

Pauku, R.L., Lowe, A. and Leakey, R.R.B. (2010) Domestication of indigenous fruit and nut trees for agroforestry in the Solomon Islands. *Forests, Trees and Livelihoods* 19, 269–287.

Shiembo, P.N., Newton, A.C. and Leakey, R.R.B. (1996a) Vegetative propagation of *Irvingia gabonensis* Baill., a West African fruit tree. *Forest Ecology and Management* 87, 185–192.

Shiembo, P.N., Newton, A.C. and Leakey, R.R.B. (1996b) Vegetative propagation of *Gnetum africanum* Welw., a leafy vegetable from West Africa. *Journal of Horticultural Science* 71, 149–155.

Shiembo, P.N., Newton, A.C. and Leakey, R.R.B. (1997) Vegetative propagation of *Ricinodendron heudelotii* (Baill) Pierre ex Pax, a West African fruit tree. *Journal of Tropical Forest Science* 9, 514–525.

Tchoundjeu, Z. and Leakey, R.R.B. (1996) Vegetative propagation of African mahogany (*Khaya ivorensis*): effects of auxin, node position, leaf area and cutting length on rooting. *New Forests* 11, 125–136.

Tchoundjeu, Z. and Leakey, R.R.B. (2000) Vegetative propagation of African mahogany: effects of stockplant flushing cycle, auxin and leaf area on carbohydrate and nutrient dynamics of cuttings. *Journal of Tropical Forest Science* 12, 77–91.

Tchoundjeu, Z. and Leakey, R.R.B. (2001) Vegetative propagation of *Lovoa trichilioides*: effects of provenance, substrate, auxins and leaf area. *Journal of Tropical Forest Science* 13, 116–129.

Tchoundjeu, Z., Avana, M.L., Simons, A.J., Assah, E., Duguma, B., et al. (2002) Vegetative propagation of *Prunus africana*: effects of rooting medium, auxin concentrations and leaf area. *Agroforestry Systems* 54, 183–192.

Case Studies from the Pacific   8

> Efforts to domesticate fruit trees should focus on integrating biological and ecological considerations with social aspirations, and should increasingly focus on addressing a variety of production systems rather than assuming one ideal typical production system. It also allows the identification of new approaches towards plant domestication in response to newly emerging social demands regarding multifunctional rather than specialized production systems.
>
> Freerk Wiersum (2008) Domestication of trees or forests: development pathways for fruit tree production in South-east Asia. In: Festus Akinnifesi *et al.* (eds) *Indigenous Fruit Trees in the Tropics: Domestication, Utilization and Commercialization*. CAB International, Wallingford, UK, pp. 70–83.

> While today's wild biodiversity is under unprecedented pressure, promising signs of innovation are coming from many parts of the world – from low-income farmers who are directly dependent on threatened wild resources, as well as from scientifically-trained agroentrepreneurs. Innovative ecoagricultural approaches can draw together the most productive elements of modern agriculture, new ecological insights, and the knowledge local people have developed from thousands of years of living among wild nature.
>
> Jeffrey McNeely and Sara Scherr (2003) *Ecoagriculture: Strategies to Feed the World and Save Biodiversity*. Island Press, Washington, DC.

In 2001, I was appointed to James Cook University (JCU), in Cairns, to set up a new research unit – the Agroforestry and Novel Crops Unit in the School of Tropical Biology. While in Australia, I was keen to work in the Pacific region and hoped that there would be interest in the domestication of indigenous fruits and nuts. I soon discovered that there are some very tasty local nuts – among them were cutnut (*Barringtonia procera* – Fig. 8.1), galip nut, ngali or nangai (*Canarium indicum* – Fig. 8.2), okari nut (*Terminalia kaernbachii* – Fig. 8.3) and Tahitian chestnut (*Inocarpus fagifer* – Fig. 8.4). Consequently, I visited several countries to discuss domestication projects.

One of the outcomes of this travel was a postgraduate training project for Richard Pauku, who came to JCU for 3 years to do his PhD, although

Fig. 8.1. Fruits and kernels of cutnut (*Barringtonia procera*) in the Solomon Islands.

his field work was done back home at Ringii Cove on Kolombangara Island, one of nearly 1000 islands forming the Solomon Islands. Many of them are beautiful atolls, with lush green vegetation surrounded by white sand and ultramarine water dotted with reefs. It looks like an idyllic paradise. However, I was aware that the country had some problems of social unrest, and indeed insurrection, violence, kidnapping and murder, as the result of ethnic clashes between people from the large and populous islands of Malaita and Guadalcanal. Two armed militia, the Isatabu Freedom Movement and the Malaita Eagles Force, led by warlords had emerged from this and the problem had spread across the country.

Richard's project involved doing a participatory priority setting with farmers to decide which two species he should study. The result of this was the choice of cutnut and Tahitian chestnut. He then collected seeds from around the island and established a simple nursery where he could develop some non-mist propagators and establish his plants. Some of these were then planted to form hedges that could be regularly cropped for cuttings. Using these plants he then started to test the basic principles of vegetative propagation in order to develop protocols for his chosen species. In the event, he was lucky that both his species were easy to root and he was quickly able to build up a large number of plants, both from juvenile cuttings and mature marcotts.

Fig. 8.2. Kernels of galip nut (*Canarium indicum*) for sale in the town market in East New Britain, Papua New Guinea.

Fig. 8.3. Fruits (a) and kernels (b) of *Terminalia kaernbachii* (okari nut) in Papua New Guinea.

Fig. 8.4. Fruit of *Inocarpus fagifer* (Tahitian chestnut) in the Solomon Islands.

In the last chapter I mentioned the need to find a way of overcoming the difficulties associated with rooting cuttings from mature trees. Richard rose to this challenge, so we tried to create conditions at the top of big cutnut trees that would result in the formation of shoots with the form and characteristics of 'juvenile' seedlings. Having achieved that, Richard then compared their rooting responses with both normal mature crown shoots from the same part of the tree crown, as well as with seedlings. If the experiment had been a complete success it would have been a breakthrough, giving us new insights into the relationship between maturation and rooting ability. However, the experiment was only partially successful in that not all of the mature shoots responded positively to the treatments. I believe that I know the reason for this. Consequently, the experiment was successful enough for me to stick my neck out and say that it is not impossible to create sexually mature shoots that have the rooting ability of seedlings. Someone else now needs to pick up this challenge and demonstrate it to the satisfaction of the scientific community.

Another component of Richard's project was to examine the tree-to-tree variability of cutnut fruits and kernels from five villages around the island, again using the experience gained in Cameroon and Nigeria. The results of this confirmed that these trees were just as variable as those in West Africa and that they had very similar patterns of variability. Richard then went on to use modern genetic techniques to assess the degree of relatedness of 120 cutnut trees from the five villages around Kolombangara Island. Using these data he was able to show that the plus-trees selected for their large kernel size were not closely related. This confirms our assumption that the level of genetic selection currently occurring in agroforestry tree domestication is unlikely to greatly reduce the species genetic diversity.

When Richard had completed his PhD he returned to the Solomon Islands to establish an NGO and start a participatory domestication programme along the lines of the Cameroon projects. After his return, I visited Honiara one more time. It was for a meeting to discuss the intellectual property issues surrounding the award of a patent to an Australian company for a medicinal product made from the kernel oil of *Canarium* nuts. During the course of a workshop we discussed the various options, which ranged from seeking to quash the patent and sue the businessman for damages, to recognizing that the patent would create business opportunities for farmers to supply the patent holder with high quality kernel oil. To my knowledge, no action has been taken as there were legal complications around the enactment of international agreements. This emphasizes how difficult it is for many countries to keep up to date with international protocols.

At more or less the same time as the Solomon Islands project, it was agreed that I should develop another project to domesticate galip nuts in the East New Britain province of Papua New Guinea in partnership with the National Agricultural Research Institute (NARI) of Papua New Guinea and Macro Consulting Group in Queensland.[1] It started with a 1-year feasibility project to convince the donor (Australian Centre for International Agricultural Research; ACIAR) that indigenous nuts were worthy of domestication and that their domestication would have sufficient impact on the economy of the country. Interestingly, archaeological evidence suggests that galip nuts have been of importance to man for 14,000 years in Papua New Guinea.

I made numerous trips to East New Britain to visit NARI's Lowlands Experimental Agricultural Station at Keravat near the town of Rabaul, which had virtually disappeared when the local volcano erupted, covering the town with ash. The first phase of the project was a feasibility study to determine if there was real potential to improve the livelihoods of local people by domesticating galip nuts. To evaluate the current situation we implemented surveys of both local producers and consumers. They ranked galip nuts as the most important nut tree species both for food and for income generation. Everybody who was interviewed confirmed that they regularly ate the kernels and that the supply of nuts did not meet the demand. Most farmers indicated that they would like to grow more of these trees for domestic consumption and for income generation. From this we concluded that there was great potential for the industry to expand sales at all levels from urban street vendors (Fig. 8.2) to supermarkets. In addition, Colin Bunt examined the potential of regional trade and the opportunities for export and value adding through the establishment of a Melanesian supply chain.

During the preliminary phase of the project we also assessed the tree-to-tree variation in fruit, nut and kernel size and did chemical analyses of the kernels, which showed that the variation in size was paralleled by variation in fat content and fatty acid profile, as well as by vitamin E, phenolic and antioxidant contents. Perhaps most interestingly, however, there was enormous tree-to-tree variation in the anti-inflammatory properties of the kernel oil (similar to the variation shown in Fig. 6.1), indicating that there would be real opportunities to select trees and domesticate them for their medicinal value. This interesting finding comes loaded with sensitivity about the use and potential abuse of traditional knowledge, as we saw in the Solomon Islands, and so has to be very carefully handled.

The outcome of our feasibility study was an agreement that ACIAR would finance a 3-year tree domestication programme aimed at enhancing the livelihoods of subsistence farmers in Papua New Guinea. The second phase of the project started before I left Cairns. It is making a much more intensive study of the variation and is starting the domestication process. Originally this was based on the Cameroon participatory domestication model, but now there is a much stronger focus on the research station pathway to domestication. This is partly due to the interests of larger-scale producers, and partly to the interests of NARI and the individuals involved.

Up to the time I returned to the UK, the biggest problem faced by this project was the poor success rates achieved by vegetative propagation. To overcome this we replaced the old mist propagators with non-mist propagators and established a stockplant garden that provided different shade and fertilizer-use regimes aimed at providing staff with a clear demonstration of the importance of stockplant management for quality. The result of this was a very substantial improvement under the best stockplant management regime from less than 10% to over 80%. Hopefully, this problem has now been solved and this knowledge can be extended to the villages for wider propagation of superior trees.

The future of galip nut looks very good, as there is a big demand locally for the nuts and the prospect of expanding this up to a regional if not international trade. Furthermore the future of cocoa in the country is threatened by the larvae of a moth (*Conopomorpha cramerella*) known as the cocoa pod borer. The effects of this pest on cocoa support the suggestion that galip nuts might be developed as a new export crop to replace cocoa. I look forward to hearing how this progresses.

Our third project in the Pacific arose from a letter I received from the Forest Department in Vanuatu, asking if we would be interested in doing a sandalwood project. The correspondence indicated there had in the past been a large sandalwood resource in Vanuatu, which was now threatened by over-exploitation. The particular species was *Santalum austrocaledonicum*, which is one of the sandalwood species found in the Pacific. The Vanuatu Department of Forests wished to revive the industry locally by promoting a planting programme to supplement the dwindling natural populations on the western sides of the islands of Espiritu Santo, Malekula, Moso (off Efate island), Erromango, Tanna and Aniwa.

The history of the commercial interest in sandalwood for its perfumed timber goes back to the early days of global exploration and the spice trade in South Asia, South-east Asia, Australia and the Pacific and led to serious deforestation and over-exploitation in the 1800s. In the mid-1800s, the principal trade of sandalwood heartwood was to China and the Far East. Sandalwood oil for the perfume industry is graded and priced depending on its content of the essential oils α- and β-santanol graded for quality on an international standard derived from Indian sandalwood (*Santalum album*). Lower quality heartwood samples are generally used for joss sticks in the incense industry, and for wood carvings, etc.

Sandalwood is an interesting, and perhaps challenging, species for domestication in that the sandalwood tree is a parasite on the roots of other plant species. The concept of domesticating a parasite is a novel idea, and of course we had no idea how much of the production of scented wood could be attributed to the genetics of the parasite itself and how much might be the influence of the genetics of the host. When we were approached by the Vanuatu Department of Forests, I instantly wondered how much tree-to-tree variation there was in the quantity and quality of the perfumed essential oils, and whether or not our approach to domestication was equally appropriate for heartwood, as we have already seen it is for fruits and nuts. Surprisingly, despite the scale, value and importance of the sandalwood trade there seemed to be very little information about the variation found within any of the sandalwood species. This therefore seemed like a very exciting project, something that would be very interesting academically, as well as very important to the industry, and so to the economy of Vanuatu.

Once we had the go-ahead to start the project we employed Dr Tony Page to implement the project and his first task was to go over to Vanuatu and start collecting heartwood samples. I joined him for the first 2 weeks, while we tested our techniques and made contacts with people in government and in the industry, especially the two oil distillers in the country. We needed to introduce Tony to all the stakeholders, as well as making contact with people who could introduce Tony to the farmers.

So as to avoid depriving the villagers of their resource we wanted to collect our heartwood samples with the minimum damage to the tree. For this purpose we purchased a drill that removed a wood core from the base of the stem. Having taken our cores, we sealed the holes and collected a lot of other descriptive data about the tree and its environment, especially making a note of the other plants growing nearby that were the likely hosts.

From our early discussions it was clear that making the stem-core collections was going to be difficult. In some areas the trees were scarce, while in others they were located in remote areas scattered over many islands, with only limited access by plane, boat or on foot. Thus developing an itinerary was quite a major logistical problem, as we did not really know where the sandalwood populations were or how easy it was going to be to get to them. So, Tony put together a team[2] to mount a genetic resources exploration expedition to locate and sample the sandalwood population across the six islands where we had been informed that it grew.

The completed collection of 264 heartwood cores came from 11 sandalwood populations on the six islands of Santo, Malekula, Moso, Erromango, Tanna and Aniwa. The core samples were then sent to Wollongbar Agricultural Institute which specializes in the analysis of essential oils. When, finally, the results arrived they were spectacular. Just as we had found in the measured traits of fruit and nut trees there was very considerable tree-to-tree variation in all four of the important essential oils (α- and β-santanol, β-curcumen-12-ol and *cis*-nuciferol), indicating that there were enormous opportunities for genetic selection. Very interestingly, the quality of the oil from the best trees was in excess of the international standards. This of course suggests that if growers in Vanuatu were to develop cultivars from these best trees they would be in a greatly improved position to market *S. austrocaledonicum* oils to the international perfume houses.

The results were also interesting for some other reasons. In addition to the chemistry, there was big tree-to-tree variation in heartwood colour, oil concentration and the amount of heartwood in the stems. Amazingly, this industry is not based on scientific knowledge of how to recognize the best trees. Instead it is based on little more than folklore. We were interested in seeing how well things like heartwood colour are related to the oil quality. The industry believes that the darker, reddish heartwoods have the highest quality. Our laboratory results indicated that heartwood colour was in fact unrelated to oil quality; although there was a weak relationship between heartwood colour and the concentration of the oil. This study also provided no evidence in support of some of the other predictors of good trees – such characteristics as the sex of the tree and the shape of the leaves. Furthermore, we found no evidence that the oil quality or yield were affected in any recognizable way by the species of the surrounding vegetation on which this parasite would be hosted.

This study in Vanuatu was virtually repeated in another sandalwood species (*Santalum lanceolatum*) found in Cape York in Far North Queensland. In this case the studies were done with the involvement of indigenous communities in the hope that any commercial developments coming from the project could be implemented by, and directly benefit, these Aboriginal communities. Interestingly, the results of the study in terms of the tree-to-tree variation in oil and heartwood characteristics were very similar to those of the Vanuatu project, in terms of both the general patterns of variation in the presence of the essential oils and the relationship of the best trees with the international standards of the perfume industry. All these results suggested that there were big opportunities in both species to move to a domestication programme. However, this was not straightforward as a parallel study of the genetic diversity in our sandalwood population by Dr Michelle Waycott showed that the trees were often closely related, especially in the smaller populations. This meant that we would need to add a cross-breeding programme to increase the genetic diversity in the populations. We therefore decided to implement two strategies running in parallel: (i) to establish seed orchards with unrelated trees, from either the same or different islands, to rescue the genetic diversity for the long-term domestication; and (ii) to establish stockplant gardens of the best trees for the shorter-term domestication by vegetative propagation.

While my job in JCU was to develop overseas projects, it was clear that there was also a need for the Agroforestry and Novel Crops Unit to play a local role within Far North Queensland. As Aboriginal Australians, like their counterparts all over the world, have used native species to meet their needs for a wide variety of food, medicines and many other day-to-day products, I hoped there would be opportunities to work with them to domesticate some indigenous food species – locally known as 'bush tucker'.

My final case study therefore comes from Australia, where the small bush tucker industry is based on a range of preserves, sauces and flavourings that are sold in tourist shops and to a lesser extent a few supermarkets. Perhaps more importantly there are also a small number of restaurants that serve bush tucker meals, mostly based on kangaroo and wallaby garnished with some of the sauces and flavourings. The food served in these restaurants is very tasty and I see no reason why with some development there should not be restaurants in every city of the world selling Australian cuisine. This might then lead to bush tucker products being sold on supermarket shelves worldwide, just as for so many other forms of cuisine. However, there was a long way to go before this would become a reality.

The Native Foods industry, as it was called, operated in an 'ad hoc' manner, without a commercial strategy, government policy and with minimal involvement of indigenous people. For example, ingredients were mostly supplied by white Australians growing indigenous fruits on their farms, although some Aboriginal communities practised wild harvesting.

My hope was to develop community domestication projects like those I have described in Africa and the Pacific and so to increase the involvement of Aboriginal communities to a much greater degree. It seemed to me that if done in a way that was sensitive to the culture and interests of the Aboriginal people it might enhance their self-respect and empower them in their quest for a better future. This might even go some way towards healing the wounds left by the anger and frustrations of past injustices, which have destroyed the trust that Aboriginals have in the government, and perhaps even lead to some reconciliation. These are lofty ambitions. The question, of course, was: how could this be done?

To explore the possibilities, I started to talk to local communities about the development and domestication of bush tucker. The idea was always received favourably and as a result I had a series of meetings with community leaders to develop project proposals. I also interacted with the Aboriginal Rainforest Council and spoke at some of their meetings and training workshops. With this interest from the Aboriginal community, I tried to identify project funds from different levels of government, and to involve local organizations with a mandate for economic development in Aboriginal communities. In developing these linkages I had great support from the staff of local development organizations,[3] but funding proved to be a severe constraint.

I soon discovered that there was one big difference between the way in which indigenous Australians and the indigenous people in Africa and the Pacific used traditional food species. In Australia, the Aboriginals were basically hunter-gatherers, although they used fire as a management tool, while in most other places that I have worked the local people have been farmers. Thinking about this it seemed to me that it was unlikely that Aboriginal Australians were going to become farmers, but if they were to develop plant nurseries in their communities they could develop cultivars protected by their rights to traditional knowledge. They would then be at the heart of the industry and could supply white farmers with their 'indigenous' products. This would have the added commercial benefit that it would capture the international interest in the Aboriginal culture of Australia.

With these thoughts in mind I tried to set up domestication projects in three Aboriginal communities with requests for funding to: (i) the Rural Industries Research and Development Centre of the Federal Government; (ii) the Indigenous Business Capacity Building Program; and (iii) the Indigenous Business Establishment Program managed by the Department of State Development and Innovation of the Queensland Government. However, this support was not forthcoming and all of these initiatives ended in failure.

With inadequate funding for the sorts of project that have been successful in other parts of the world, the best we were able to achieve was a training project for indigenous students at the Technical and Further Education College at Innisfail[4] in partnership with the Ma:Mu community.[5] This was funded by the Rainforest Cooperative Research Centre and taught students the basic theory and practice of vegetative propagation and tree selection techniques in tree domestication. Through the work of the students the college developed a large stock of plants of many different species.

The plan was then to plant these on the land of the Ma:Mu community, but unfortunately this did not happen before Cyclone Larry hit Innisfail.

To try to get important messages through to government and to raise awareness of the issues around the possible development of a bush tucker industry we organized a Bush Tucker Summit[6] under the aegis of the Network for Sustainable and Diversified Agriculture. This was attended by 78 delegates from State Government, the business community (food companies, restaurants, agribusiness, farmers and producers, trade associations, etc., including some bush tucker enterprises), research and education communities and, critically, a number of different Aboriginal communities (Djabugay, Ma:Mu, Yarraba, as well as communities from Cape York in Lockhart River, Alice Springs, Napranum, Pompuraaw and Mapoon). The purposes of this summit were: (i) to bring together participants who could help foster an attitude of 'mutual support' among all industry stakeholders; (ii) to identify the interest in and constraints to the development of the bush tucker industry; and then (iii) to chart a path and strategy for the development of the industry.

The summit paid particular attention to: (i) the constraints to marketing; (ii) supply chain management; (iii) value adding; (iv) opportunities for species domestication; and (v) the protection of intellectual property rights and traditional knowledge. It was foreseen that the summit should also raise political awareness of the potential of the industry. Seith Fourmile, who welcomed the delegates on behalf of the Traditional Owners of the Country, confirmed the interest of many indigenous Australians in developing a bush tucker industry that would result in job creation and sustainable economic benefits and emphasized the need to ensure that indigenous peoples weren't disadvantaged by, or excluded from, the industry development process, as had often been the case with commercialization of indigenous art and other forms of traditional knowledge.

Key among the outputs of this Summit was the indication that there are already active and interested parties among both the Producers (growers and plant nurseries) and the various levels of Consumers (the food service sector, the retail sector and exporters), who are trying to promote the industry and that indigenous Australians have great interest in playing their role. Nevertheless, there is much to be done to achieve a viable bush tucker industry. There was consensus that there needs to be equitable involvement of all stakeholders, active collaboration between Indigenous and non-Indigenous people, and recognition and protection of Indigenous rights.

Additionally, it was recommended that the industry requires a Strategic Plan and a 'Peak Body' to: define market opportunities; foster the development of the different players in the industry; promote the transfer and use of information; encourage a review of policy and legislation impeding the development of the industry; and promote business, technical and marketing training, and policy support at the regional and national levels.

I understand that since my retirement the Peak Body has been established and that progress continues to be made. This is encouraging. However, the net effect of my personal efforts to get some movement in the bush tucker industry in Australia came to very little, although they did at least raise awareness of some of the issues and challenges. I believe this case study therefore identifies important social issues affecting the chances of involving indigenous people in the creation of new crops from indigenous food plants. In a developing country like Cameroon or the Solomon Islands, it seems to me incentives to get involved are greater due to the need to feed the family and survive hard times. This may be because there are no government-funded social services to provide financial support when times are challenging.

Notes

[1] Dr John Moxon, Mark Johnston, Tio Nevenimo and Colin Bunt.
[2] Hanington Tate, Chanel Sam, Joseph Tungon, Philimon Ala, Leimon Kalomor of Vanuatu Forest Department, Ken Robson, Geoff Dickinson and David Osborne of the Australian Department of Primary Industries and Fisheries.
[3] Tom Viera of Far North Queensland Area Consultative Committee, Tracy Scott-Rimington, Steve Oldham, Fred Marchant and Nola Craig of Cairns Regional Economic Development Corporation, Sue Fairley an AusIndustry Small Business advisor, Mark Annandale of Department of State Development, and others.
[4] Richard McIntosh, Marianne Helling, Rob Tranent and Kylie Freebody.
[5] The Chairman was Victor Maund.
[6] I was supported by Colin Bunt and Colin Taylor.

Further Reading

Asaah, E.K., Tchoundjeu, Z., Leakey, R.R.B., Takousting, B., Njong, J., *et al.* (2011) Trees, agroforestry and multifunctional agriculture in Cameroon. *International Journal of Agricultural Sustainability* 9, 110–119.

Bunt, C. and Leakey, R.R.B. (2008) Domestication potential and marketing of *Canarium indicum* nuts in the Pacific: commercialization and market development. *Forests, Trees and Livelihoods* 18, 271–289.

Leakey, R.R.B. (2006) Traditional trees – a key to well-being and prosperity. In: Elevitch, C.R. (ed.) *Traditional Trees of Pacific Islands: Their Culture, Environment and Use.* Permanent Agriculture Resource, Holualoa, Hawaii, pp. ix–xii.

Leakey, R.R.B. (2012) Participatory domestication of indigenous fruit and nut trees: new crops for sustainable agriculture in developing countries. In: Gepts, P., Famula, T.R., Bettinger, R.L., Brush, S.B., Damania, A.B., McGuire, P.E. and Qualset, C.O. (eds) *Biodiversity in Agriculture: Domestication, Evolution and Sustainability.* Cambridge University Press, New York, pp. 479–501.

Leakey, R.R.B. and Cornelius, J. (2010) Special issue on agroforestry tree domestication. *Forests, Trees and Livelihoods* 19, 199–316.

Leakey, R.R.B., Fuller, S., Treloar, T., Stevenson, L., Hunter, D., *et al.* (2008) Characterization of tree-to-tree variation in morphological, nutritional and chemical properties of *Canarium indicum* nuts. *Agroforestry Systems* 73, 77–87.

Leakey, R.R.B., Weber, J.C., Page, T., Cornelius, J.P., Akinnifesi, F.K., *et al.* (in press) Tree domestication in agroforestry: progress in the second decade (2003–2012). In: Nair, P.K. and Garrity, D. (eds) *Agroforestry: The Way Forward.* Springer Verlag, Dordrecht, The Netherlands.

Nevenimo, T., Moxon, J., Wemin, J., Johnston, M., Bunt, C., *et al.* (2007) Domestication potential and marketing of *Canarium indicum* nuts in the Pacific: 1. A literature review. *Agroforestry Systems* 69, 117–134.

Nevenimo, T., Johnston, M., Binifa, J., Gwabu, C., Anjen, J., *et al.* (2008) Domestication potential and marketing of *Canarium indicum* nuts in the Pacific: producer and consumer surveys in Papua New Guinea (East New Britain). *Forests, Trees and Livelihoods* 18, 253–269.

Page, T., Southwell, I., Russell, M., Tate, H., Tungon, J., *et al.* (2010) Geographic and phenotypic variation in heartwood and essential oil characters in natural populations of *Santalum austrocaledonicum* in Vanuatu. *Chemistry and Biodiversity* 7, 1990–2006.

Pauku, R.L., Lowe, A. and Leakey, R.R.B. (2010) Domestication of indigenous fruit and nut trees for agroforestry in the Solomon Islands. *Forests, Trees and Livelihoods* 19, 269–287.

Marketing Tree Products 9

> There is ample evidence that the results of international agricultural R&D bring substantial returns to the agricultural sectors of developed, as well as developing, countries which, in total, far exceed each year the sums allocated by developed countries to these activities. Yet these returns still do not constitute the major benefits. In the medium to long term, there are also major gains from the effects of agricultural development on world trade.
> Derek Tribe (1994) *Feeding and Greening the World: the Role of International Agricultural Research.* CAB International, Wallingford, UK.

> Targeting market and trade policies to enhance the ability of agricultural and systems of agricultural science and technology to drive development, strengthen food security, maximize environmental sustainability, and help make the small-scale farm sector profitable to spearhead poverty reduction is an immediate challenge around the world.
> IAASTD (2009) *Agriculture at a Crossroads: Synthesis Report* (edited by Beverley McIntyre, *et al.*). Island Press, Washington, DC.

When our current food crops set out on the road to domestication, the aim was enhanced domestic consumption. Three processes commonly emerged: (i) deliberate cultivation; (ii) genetic improvement; and (iii) extending the benefits of the harvest to other people through bartering, commercialization and trade. Thus domestication and commercialization go hand in hand. This critical relationship sometimes gets forgotten. I hope that it is obvious that there is absolutely no point in developing a new crop if the farmers are not going to be able to sell their products. Conversely, and equally important, there is no point in developing new markets for products which cannot be supplied regularly or for which the product does not meet expected product uniformity and quality. It is for all these reasons that the ICRAF initiative for the Trees of Life follows a strategy of being 'market led and farmer driven'.

Typically, as we have seen in earlier chapters, products harvested from individual wild trees are highly variable from tree to tree. The impact of this variability was clearly illustrated by a market study of safou in

Cameroon, which showed that wholesale traders did not pay farmers high prices for loads of mixed-tree fruits, while small-scale market retailers selling the products of individual trees charged consumers high prices for large and tasty fruits and very low prices for small and less desirable fruits. This is one of the reasons for engaging in tree domestication aimed at the mass production of a much more uniform and superior product – often a named variety. Hopefully wholesalers will be prepared to pay more when they can buy a full load of superior fruits from domesticated cultivars. This is standard practice in industrialized countries where there would not be much demand for a kilo of crab apples, but this practice needs to be initiated for indigenous fruits being traded in local markets of developing countries.

Another aspect of this issue is the short shelf life of many fruits. Domestication can help here, as it is possible to develop cultivars with different fruiting seasons. For example, in Cameroon, some safou trees are either 'early' or 'late' producers and a few are productive outside the normal season, like the cultivar 'Noël', which fruits in December when the main season is May–October.

A second way to reduce the seasonality of the market and so to expand the market potential of fruits and other products is to engage local entrepreneurs in processing and simple storage (bottling, canning, drying, freezing, etc.) techniques. Other approaches to value adding include the processing of paste, baking of biscuits, etc., which create a year-round supply of preserved or processed products. To date this work is still in its infancy in agroforestry, but there are examples of local entrepreneurs drying fruits and nuts in Cameroon, Vanuatu and the Solomon Islands, while in southern Africa communities are engaged in jam making and the brewing of alcoholic beverages. At best, however, this could be described as a 'cottage industry', but this of course has the potential to expand and create business and employment opportunities in the communities. This aligns work to extending the shelf life of fruits and nuts to the overall poverty alleviation strategy of agroforestry, as value adding is one of the best ways of increasing the market price. However, to maximize the opportunities for poverty alleviation it is important that the methods of preservation are relevant to the situations of community members. This hopefully is not too difficult for local and perhaps regional markets, but does become much more complex if the products have export potential to Europe or the USA where the food standards legislation requirements are much more rigid.

Unfortunately, in our exploitative modern world, the commercialization of anything sets '$ signs' spinning in the eyes of business people, and all too often this is the signal for exploitation. When it comes to exploitation of the poor, vulnerable and most marginalized people in the world, one would hope that sympathy and understanding would be the dominant response, but sadly some see this as an opportunity – a chance to profit because the poor have little, if any, capacity to stand up to their oppressors. What can be done about this?

I think we need to examine the issues at three levels. Before looking at them, however, we need to remind ourselves that we are talking about farmers who are using their traditional and local knowledge, together with newly learnt domestication skills, to produce cultivars developed from local 'plus-trees'. As we will see in Chapter 11, small community programmes in Cameroon have brought well over 2000 mature plus-trees into their village nurseries for further multiplication by vegetative propagation. Some of these they will cultivate on their own farms to produce products for domestic consumption and to sell in local markets. Others will be multiplied up as planting stock to sell to other farmers and to start the long haul up an income-generation pathway. For them this is a life-transforming opportunity, not to become wealthy, but to escape poverty.[1]

The first issue to address is that these developments are moving forward more rapidly than the international efforts to develop farmer and community rights over both their traditional knowledge and their investments in time and effort. In international law there are four main ways in which innovators can claim rights to their 'inventions' and commercial investments. These are: (i) patents; (ii) trademarks; (iii) plant breeders' rights; and (iv) copyright. Of these, plant breeders' rights are the most relevant to crop domestication activities. However, these forms of intellectual property protection have been developed to meet the complex legal needs of large commercial plant-breeding companies, and are not appropriate for poor people.

To try to redress this deficiency, there have been years of discussion in international meetings, but to date nothing has been formally recognized. If protection of these rights is not achieved very soon, there is the danger that other people will reap the benefits of the pioneering work being done by villagers. Protection of this kind must therefore be achieved if participatory domestication is to be part of the way that poor farmers can lift themselves out of poverty. In the jargon, this is about 'access and benefit sharing' or, in plain language, making sure that the person making the innovations gets a fair share of the economic benefits. The development of tree cultivars by indigenous peoples is one aspect of this.

The absence of a fair way to reward local people for the loss of their traditional knowledge and culture is one of the reasons that billions of people in remote corners of the world are in poverty. Theft of this sort is known as 'biopiracy' and it is no longer acceptable; we have to find ways to ensure that indigenous peoples are properly rewarded if they agree to their traditional knowledge being commercialized. Books have been written about this and I am not going to try to paraphrase them. All that I will say is that I am helping ICRAF to develop an interim measure that hopefully will clearly record who has developed which cultivar, where and when it was done, what is unique about it, and formally lodge this information together with a genetic 'fingerprint' of the mother tree. Hopefully this will mean that in the event of someone misappropriating an important cultivar and marketing it as their own, it will be possible to mount a legal challenge for compensation on behalf of the original farmer/community.

The second issue that has to be addressed is that when communities are trading the products from domesticated cultivars they are paid a fair price. Already there is an ever-increasing array of processed products from wild, or semi-domesticated, species appearing in international trade (Fig. 9.1). This too is an important development, but comes with the risk that producers will be unfairly exploited. To minimize this risk, organizations like PhytoTrade Africa work with producer communities to help them to develop strong and viable trade associations based on forward-thinking and market-oriented agreements with commercial companies. This is necessary to ensure they have long-term access to formal markets and can be competitive over the long term as suppliers in the value chain.

On one trip to South Africa, we thought about these issues while sitting on a bench in the hot sun outside the hut of a local farmer, drinking traditionally brewed marula beer. Actually, I think it is more like a cider, but it is very tasty and very refreshing, especially in a hot and dry climate. It would be a very popular drink in many places around the world, but before this could be commercialized it would be very important to ensure that 'benefits' flowed back to the local people as the drink would be based on their traditional knowledge. In addition, local people should have the opportunity to veto the commercialization of such products if they feel it threatens their culture and tradition.

The third issue to address is that there has in the past been no recognition that a species has a legal area of origin. As a result we have well-recognized crops that are developed and marketed well outside their geographic area of origin, even in different continents, making huge profits for international companies, often leaving the people of the area of origin with nothing. For example, the Trees of Life initiative to domesticate species as new crops has the express intention of creating a pathway out of poverty for about half of the world's population. It would be grossly unfair and an unmitigated disaster if a company were allowed to develop monocultural plantations of the new crop in some overseas location with a similar climate and better access to markets. At the moment there is nothing in international law to prevent this from happening. It will also lead to distorted market forces that increase the inequity between the rich and the poor. Ways have to be found to avoid the problem. In this regard, it is highly relevant and extremely encouraging that Unilever plc has made it clear that it is helping African communities to develop a new oil-tree crop from *Allanblackia* species as a new African commodity.

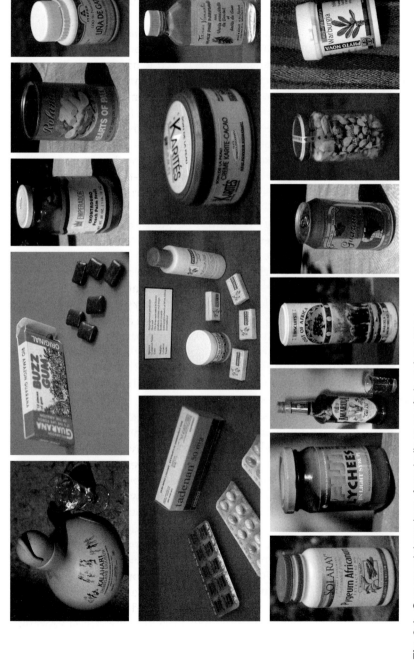

Fig. 9.1. Commercial products from indigenous fruit and nut trees.

In addition to these three policy approaches to improving the lot of the poor farmer producing indigenous tree cultivars, there are some natural factors on the side of the smallholder practising mixed cropping with species with predominantly local market potential. For example, in contrast to large-scale plantations of indigenous species grown in their native environment, small-scale mixed cropping systems are less likely to be vulnerable to an explosion of predators, weeds and diseases that would require the additional expense of pesticides. This is due to the greater chance that these outbreaks will be checked by the activation of the natural checks and balances in the agroecosystem imposed by parasites, predators and defoliators as we discussed in Chapter 4.

Fortunately for the Trees of Life initiative, the economic laws of 'supply and demand' do not favour the large-scale marketing of products with relatively small-scale markets. At present most indigenous tropical fruits and nuts have a small value in formal trade. For example, in 2001, the annual trade of the products of five key species in Cameroon was valued at US$7.5 million, of which exports generated US$2.5 million. One strategy to minimize the risk of market domination by overseas big business is, therefore, to support the domestication of large numbers of tree species producing AFTPs, especially those with local and regional market potential. In this way, if even a few do become global commodities, there will still be plenty of others with market potential limited to local and regional trade. Again the Unilever development of *Allanblackia* oil is encouraging, as they have developed a business paradigm in support of community-oriented agroforestry production. If this new paradigm is successful then we can perhaps have the benefits of large markets with potential to support large numbers of poor farmers, without the risks of their initiatives being undermined by big business.

Despite the above points, there is no getting away from the fact that markets are essential for smallholder households to increase their standard of living, and expanded market opportunities could lead to their exploitation by businessmen. Thus it is clear that commercialization is both necessary and potentially harmful to small-scale farmers practising agroforestry. Great care must therefore be taken to ensure that the positive benefits are not overwhelmed by the negative impacts.

Meaningful discussion of the pros and cons of expanding the commercialization of indigenous tree products vis-à-vis their ability to benefit local rural communities is often constrained by credible information. While I was working at JCU in Australia, I was actively involved in a multi-institutional, multidisciplinary 'Winners and Losers' project funded by UK DFID to generate useful information about the impacts of domestication and commercialization of AFTPs, and specifically marula (*Sclerocarya birrea*) in South Africa and Namibia.[2]

Marula trees occur in woodland savannah throughout dry Africa, but in this project we were only concerned with the subspecies found in the Miombo woodlands of southern Africa. The fruits are very tasty, with a slightly 'sharp' tang and as we have already seen they can be made into a very refreshing drink. Elephants are also said to be very keen on these fruits and there are stories of elephants eating fallen and fermented fruits and getting drunk. People too like the fermented fruits and, as mentioned earlier, the traditional 'beer' is important in a number of cultural ceremonies. Juice of the marula fruit is also used as the flavouring of a cream liqueur called 'Amarula', which is manufactured by Distell Corporation. This is being marketed internationally and can be bought in supermarkets, off-licence stores and duty-free shops at airports around the world. The fruit contains a very hard nut, which in turn contains several smallish kernels. These kernels are edible and extracted as a snack food. They are also said to have aphrodisiac properties. When the kernel oil is extracted it is used as high quality cooking oil, with the unusual property of not going rancid. All of this means that there are potentially many commercial opportunities for the marula that could be enhanced by genetic selection and domestication.

Our research plan was to investigate and compare the typical 'top-down' approach to commercialization, with what could be thought of as the 'bottom-up' approach that would emerge from community domestication and marketing of marula products. Our aim was to see if certain models of commercialization would lead to negative impacts on local people. For example, without the model of participatory domestication, many organizations would think that a new fruit-tree crop should be domesticated for growth in large-scale monocultural orchards implemented by rich entrepreneurs. The only benefits that might accrue to local people from this approach would probably be employment as poorly paid labourers. Furthermore, commercialization in these circumstances might deprive local people of a culturally important and highly nutritious food.

In summary, this study highlighted that there are numerous factors that determine whether the impacts of commercialization are beneficial to local people or not (see Table 9.1). Without going into detail, the study found that the bottom-up approach had many benefits but, interestingly, it also became clear that the top-down approach could be done in ways that are sympathetic to the cultural and economic needs of local people. Indeed the model implemented by Distell Corporation had many positive attributes, although of course there is always room for improvement. In 2002, the company sourced about 2000 t of marula fruits from local people rather than supporting large-scale plantations established by expatriates. Thus, it was recognized that positive outcomes could be attained even in top-down commercialization, if the importance of community involvement

Table 9.1. The social, marketing and natural resource qualities that determine whether the impacts of commercializing indigenous fruits and nuts are positive or negative. (Source: Leakey *et al.*, 2005. Reproduced with permission from Taylor & Francis Ltd, see www.tandfonline.com.)

Winner qualities	Loser qualities
In individuals, households and enterprises	
Individuals organized as a group	Poorly organized group structure
Well informed about markets	Poorly informed of markets
Good access to transport	Poor access to transport
Coordinated production	Uncoordinated production
Small 'input cost: revenue received' ratio	Large 'input cost: revenue received' ratio
Consistently good quality products	Variable quality products
Skilled in bargaining	Unskilled in bargaining
Well networked with good partnerships	Poorly networked
Easy and equitable access to resource	Uncertain and restricted access to resource
Fits with other livelihood strategies and sociocultural norms	Competes with other livelihood strategies and sociocultural norms
In product marketing	
Commercial opportunities	Undeveloped/poor market interest
Diversity of end markets	Limited markets
Diversity of end products	Fad or single niche products
Positive marketing image	No or negative marketing image
Unique characteristics of product	Many other substitutes
Raw product quality well matched to market	Raw product requires processing
Many buyers of raw materials and products	A monopsony – only one buyer of raw materials
Many sellers of raw materials and products	A monopoly – only one seller
Buyers aware of product or brand	Buyers ignorant of product or brand
In the tree resource	
Abundant resource	Rare resource
Plant part used is readily renewable	Slow replacement of harvested product
Harvesting does not destroy the plant	Destructive and damaging harvesting
Easily propagated	Difficult to propagate
Genetically diverse with potential for domestication	Genetically uniform or little potential for selection
Multiple uses for products	Narrow use options
High yield of high quality product	Low yielding and/or poor quality product
Valuable product	Low value product
Consistent and reliable yield from year to year	Inconsistent and unpredictable production

Continued

Table 9.1. Continued.

Winner qualities	Loser qualities
Already cultivated within farming system	Wild resource that is difficult to cultivate
Already being domesticated by local farmers	Totally wild resource
Fast growing	Slow growing
Short time to production of product	Long time to production
Compatible with agroforestry land uses	Competitive with crops; labour intensive, etc.
Hardy	Sensitive to adverse environmental conditions
Widely distributed	Only locally distributed

is appreciated and if the communities themselves work together and use their own strengths to manage and use their resources effectively.

These results also provide endorsement for the approach being fostered through community engagement in participatory domestication. Nevertheless, as already mentioned above, there is the need in the case of marula to resolve the current difficulties facing farmers wishing to protect their rights to the cultivars they produce, if we are to ensure they are 'winners'. In addition, it is clear that there is typically a need to improve product quality and yield, product processing and marketing to diversify opportunities for trade, promote equitable benefit sharing and trading partnerships. As we have seen in this chapter, these are all issues that are part of the current approach to agroforestry tree domestication.

As part of my fieldwork in South Africa, I visited the Mhala Development Centre, a project of the Mineworker's Development Agency. Here I watched women separating the pulp and juice from the nuts for sale to fruit-juice manufacturers. They also crack the nuts and squeeze oil from the kernels for sale to cosmetics companies. I found that these women were bringing in all but the very smallest fruits for processing. It was great to see these women generating income for their communities in this way. Of course, it would be even better if they were domesticating their trees and improving the quality and yield of their trees.

Before moving on, I want to highlight two elements of my contribution to this study. I assessed the tree-to-tree variation in marula fruits and nuts to assess the potential for domestication in support of commercialization. In most respects my results were similar to those already presented for other species (e.g. Fig. 6.1), so I will not repeat them here. However, at one of the farms we visited we collected fruits that were very much bigger than any I had seen elsewhere. The average fruit weight was 70 g, while those of other trees averaged 23 g (ranging from 12 to 43 g). We called this tree the 'Namibian Wonder'. Very rare individuals like this are obviously the 'Miss World' of the marula species. Trees like this are the property of the land owner, so if cultivars are developed from them the owner should be the recipient of a fair proportion of the benefits. This is relatively easy if the owner is the person developing the cultivar, but if he/she is not, then the domestication needs to be done under the terms of an 'access and benefit sharing' agreement. This brings us back to the earlier discussion of the need for progress in the development of international law.

My results also showed that, as in safou and bush mango in West Africa, some domestication has already been done by the farmers – by what I earlier called commensal domestication. This evidence strengthens the legal arguments with regard to the farmers' ownership of rights to these fruits. These arguments were further strengthened by results from Professor Charlie Shackleton, another member of the project team, who found that trees in farmers' fields were giving yields fivefold greater than wild trees. Interestingly, my own work showed that fruits from some trees in farmers' fields had bigger fruits than those in either communal land or natural woodland, again supporting the argument that farmers have been engaged in tree selection.

I also found that marula fruits varied in the number of kernels present in the nut, but this seemed to be attributable to the pollination success and not to the genetics of the tree, as fruits on the same tree had very different numbers of kernels. The explanation is that marula trees are either male or female, and not a mixture of both sexes like most species, thus if farmers cut down those trees that are not productive (as they do!), there can be insufficient males in the population to be the source of pollen. As a result the seeds – the kernels – are not formed in the nut. Clearly, this requires a remedy. One approach would be to explain to farmers about the need for a good ratio between male and female trees in their fields. Another would be to encourage the placement of beehives in the trees – so creating another source of farm income.

In this chapter we have seen that domestication and commercialization should occur simultaneously to maximize the benefits from AFTPs. While much can be achieved in relatively informal local and regional markets there is also the possibility of reaping additional benefits from the improved product quality and uniformity by processing and value adding in ways that make the products international commodities. This, however, requires the appropriate involvement of the food, pharmaceutical and other industries. We've already seen the need for trade agreements in the commercialization, but the involvement of industry in the domestication process is also needed in order to identify the product traits and 'ideotype' that will determine market acceptability. So far in agroforestry there are few examples of industry partnerships. However, I have had some involvement with two examples of public–private partnerships (PPPs) that illustrate the potential for wider collaboration between agroforestry research scientists, industry and farmer community groups.

My first experience of a PPP was during a visit to Belém, in Brazil, for a conference organized by Daimler AG and the University of Hohenheim's Institute of Plant Production and Agroecology in the Tropics and Subtropics. At first sight there would seem to be little in common between building top-of-the-range deluxe cars and either agroforestry or poverty alleviation, but at that time new and innovative ideas were coming to fruition. Some years earlier the company had come to the conclusion that they should contribute to change by supporting basic and applied research on the use of renewable, natural materials in automobile manufacturing. They established a partnership between an NGO ('Poverty and Environment in the Amazon' – POEMA), the Federal University of Pará at Belém in Brazil, and a cooperative called PRONAMAZON. At the university, basic research examined the suitability of a range of products, such as fibres, dyes, oils and rubber from Amazonian plants for industrial uses. The outcome was an eco-composite product made as a polypropylene from natural products to be fitted in the C-Class Mercedes-Benz cars manufactured in Brazil. This alternative to fibreglass was then used to manufacture the panelling of vehicles.[3] POEMA, on the other hand, worked in an integrative and participatory approach to agroforestry with communities near Belém to produce the raw materials for the eco-composite. The agroforests produced: (i) coconut, jute, sisal, curaná (*Ananas erectifolius*) and ramie (*Boehmeria nivea*) for fibres; (ii) castor oil, rubber, cashew oil, andiroba (*Carapa guianensis*) and indigo (*Indigofera arrecta*) for the factory; and (iii) a wide range of food crops and indigenous trees products of household importance for the farmers. In addition, the company assisted the participating communities by providing facilities for clean water, health and education.

This small-scale production of raw materials for car manufacturing in the rural community was a step towards improved livelihoods and greater sustainability in rural communities. However, this project was not just altruism by a large multinational company, as the use of natural products provided commercial benefits for the company. In addition, there were environmental benefits, such as lower energy requirements and reduced dependence on fossil resources for fibreglass, due to the switch to renewable resources for the eco-composite. The result of this has been that the Mercedes C-Class cars from the Brazil factory contain many natural product components that come from both environmentally and socially sustainable farming systems (Fig. 9.2) – and there is potential to expand this to trucks, trains and aircraft manufacturing. Thus we can see that the decision by this large multinational company to promote the production of raw materials through small-scale agroforestry, rather than large-scale monocultures, is particularly significant in the 'development' context.

As this PPP has matured POEMA has become a new spin-off company, POEMAtec. It now functions in a free market and is successfully delivering natural fibre products for automotive and non-automotive applications. The initiative is prospering and Daimler is continuing to

Fig. 9.2. Daimler car parts made from natural products grown in agroforestry systems by local communities. (Source: Daimler AG, 1996.)

increase the quantity of sustainable materials used in their manufacturing processes. They now concentrate on collaboration with their independent suppliers across the whole supply chain and the suppliers have taken over responsibility for ensuring that the natural products are produced sustainably. Daimler were recently acknowledged for their 'green leadership' by the award of an environmental certificate for the Mercedes-Benz S-Class in accordance with the ISO (International Organization for Standardization) design for Environment Standard 14062 in 2005. This was the world's first automobile to receive this award. In 2007, environmental certificates were also awarded to the C-Class sedan and station wagon.

I have already mentioned my second example of a PPP. It is the Novella Partnership, which involves Unilever plc, two national trading companies ('Novel Ghana' and 'Novel Tanzania'), the World Agroforestry Centre (ICRAF), the International Union for the Conservation of Nature, the Netherlands Development Organization and the Forestry Research Institute of Ghana with local communities and farmers and some NGOs. Under this partnership, the 'Novel' companies have been established in producer countries to improve the supply chain and to undertake local research. Thus the national forestry research institutes of Ghana, Nigeria and Tanzania are working with ICRAF to domesticate some of the nine tree species in the genus *Allanblackia* together with local farmers.

The commercial interest in the *Allanblackia* species arises from the discovery of new uses for the kernel oil, which is very rich in triglycerides of stearic and oleic fatty acids, has a high melting point and good structuring properties, making it an interesting ingredient for use in the manufacture of food and non-food products. The current commercial interest in *Allanblackia* is focused on margarine production as it provides structure and product stability. Compared with the conventional oils produced from palm oil and palm kernel, the use of *Allanblackia* improves the fat composition of the margarine by decreasing the level of saturated fats and increasing the level of monounsaturated fats. Again the innovative aspect of this business initiative by a multinational company is its commitment to work in partnership with African smallholder farmers to reduce poverty and support more sustainable farming practices. To stimulate the initiative Unilever has created a guaranteed market for the product. In addition, the Novella Partnership has set a moratorium on the release of germplasm from Africa. Hopefully, this pioneering project will become a model for sustainable agricultural development as it has recently been awarded the Union for Ethical Biotrade Award 2010.

I am proud that four of my former postgraduate students, Daniel Ofori, Theresa Peprah, Zac Tchoundjeu and Alain Atangana, are involved in the research to domesticate these *Allanblackia* species (Fig. 9.3), which is proving to be quite challenging. Like the other species we considered in Chapters 6 and 8, there is very considerable tree-to-tree variation in all fruit, nut, kernel and oil characteristics, so there is plenty of opportunity for the development of high quality cultivars.

I hope other industries will follow suit in genuine partnerships with local communities that lead to meaningful livelihood benefits for the communities. In this connection, we saw earlier the pioneering work

Fig. 9.3. Fruits of *Allanblackia stuhlmannii* in Tanzania.

of PhytoTrade Africa in developing carefully constructed supply chain agreements between community producers, traders and manufacturing companies. In summary, therefore, there have been some very positive outcomes in recent years from the innovative involvement of international companies in agroforestry.

Are we starting to see a new relationship between multinational companies and communities in developing countries? If so, then I think we are starting to see what Professor Jeffery Sachs, the Special Advisor to Kofi Annan when he was UN Secretary General, has called 'enlightened globalization'.[4] He proposed this concept to address 'the needs of the poorest of the poor, the global environment, and the spread of democracy'. Within this concept, international agencies and countries of the industrial north would work with partners in the south to develop new processing industries focused on the needs of local people in developing countries. Enlightened globalization is also aimed at helping poor countries to gain access to the markets of richer countries.

Over the last 60 years, agricultural intensification has run in parallel with 'globalization', but currently this approach to world trade is not really targeted at the development of an integrated global economy, in which people from all around the world work together to benefit each other. It is instead a rather USA/Euro-centric concept of global trade, with a hefty dose of self-interest, which has resulted in poor farmers in poor countries being marginalized. There is some evidence that the current global 'food crisis' is in fact linked to revolution, terrorism, etc. Thus it seems a failure to develop a more equitable approach to global trade is piling up more and more trouble for the future and so perhaps the Trees of Life initiatives are a pointer to a more promising future.

If we try to look at the big picture, it seems to me that over the past 50–60 years 'globalization' has been dominant at the expense of localization – the grassroots pathway to development relevant to local communities. However, in the last decade or two there have been a number of initiatives that have started to redress the balance between globalization and localization. For example the recognition of participatory management of natural resources and the rise of 'fair trade' and 'certification' schemes for sustainable production.

Now, with the emergence of the PPPs and the concept of 'enlightened globalization', we are beginning to see a very positive cross-fertilization of the localization initiatives in developing countries by globalized big business. This, I believe, is a new and exciting development involving diverse stakeholder groups at the local level, especially the farmers. I hope it leads to further support for sustainable production, tree domestication and the in-country processing and value adding of agroforestry products. In the next chapter we will take this idea forward. I hope that the commercial opportunities arising from a partnership that brings many of the 3 billion poor people into the global market could be highly attractive to commercial companies producing a wide range of products from raw materials grown in developing countries.

The public at large is generally aware of the food crisis, but probably very poorly informed about the deeper issues behind agriculture. There is therefore a need to raise awareness. Maybe we need an awards show for responsible living – an environment/social Oscars! In earlier generations, the rich and famous were often the people who led industrial growth and social reform. In recent years some current celebrities have similarly embraced actions against deforestation, the reduction of diseases such as malaria and HIV, and become ambassadors for change with regard to some social and environmental issues. I believe we need an initiative to promote more sustainable living – starting with agriculture.

Notes

[1] Just as a reminder, we are identifying poverty as living on less than US$1/day.

[2] In South Africa the team led by Dr Caroline Sullivan included: Drs Sheona and Charlie Shackleton, Philippa Emmanuel, Sibongile Ndlovu and Jenny Botha of Rhodes University, Grahamstown; Myles Mander, Tania McHardy, Jill Cribbins and Fonda Lewis of the Institute of Natural Resources, Scotsville; Rachel Wynberg, a South African postgraduate student of the University of Strathclyde, UK; and Adrian Combrinck, Jillian Muller, Thiambi Netshiluvhi of CSIR Environmentek, Pretoria. The team in Namibia was: Pierre du Plessis, Saskia den Adel, Andy Botelle, Cyril Lombard, Risto Laamanen and Kris Pate of CRIAA SA-DC, Windhoek. Other members of the team were: Dr Tony Cunningham of Ethnoecology Services, Fremantle, Australia and Dr Sarah Laird of the Department of Anthropology, University College, London. Dermot O'Regan and Uffe Poulsen assisted with project administration.

[3] Leihner, D.E. and Mitschein, T.A. (eds) (1998) *A Third Millennium for Humanity? The Search for Paths of Sustainable Development.* Peter Lang, Frankfurt, Germany. This includes the following three chapters: (i) Panik, F. The use of biodiversity and implications for industrial production, pp. 59–73; (ii) Mitschein, T.A. and Miranda, P.S. POEMA: a proposal for sustainable development in Amazonia, pp. 329–366; and (iii) Kübler, E. Use of natural fibres as reinforcement in composites for vehicles: research results and experiences, pp. 393–402.

[4] Sachs, J. (2005) *The End of Poverty: How We Can Make It Happen In Our Lifetime.* Penguin Books, London.

Further Reading

Beattie, A.J., Barthlott, W., Elisabetsky, E., Farrel, R., Kheng, C.T., et al. (2005) New products and industries from biodiversity. In: Hassan, R., Scholes, R. and Ash, N. (eds) *Ecosystems and Human Well-Being:* Vol 1. *Current State and Trends.* Findings of the Condition and Trends Working Group of the Millennium Ecosystem Assessment. Island Press, Washington, DC, pp. 271–296.

Leakey, R.R.B. (2007) Domestication and marketing of novel crops. In: Scherr, S.J. and McNeely, J.A. (eds) *Farming with Nature: the Science and Practice of Ecoagriculture*. Island Press, Washington, DC, pp. 83–102.

Leakey, R.R.B. and Izac, A.-M. (1996) Linkages between domestication and commercialization of non-timber forest products: implications for agroforestry. In: Leakey, R.R.B., Temu, A.B., Melnyk, M. and Vantomme, P. (eds) *Domestication and Commercialization of Non-timber Forest Products*. Non-Wood Forest Products No. 9. Food and Agriculture Organization, Rome, pp. 1–7.

Leakey, R.R.B., Temu, A.B., Melnyk, M. and Vantomme, P. (eds) (1996) *Domestication and Commercialization of Non-Timber Forest Products for Agroforestry*. Non-Wood Forest Products No. 9. Food and Agriculture Organization, Rome.

Leakey, R.R.B., Tchoundjeu, Z., Schreckenberg, K., Shackleton, S. and Shackleton, C. (2005) Agroforestry tree products (AFTPs): targeting poverty reduction and enhanced livelihoods. *International Journal of Agricultural Sustainability* 3, 1–23.

Leakey, R.R.B., Tchoundjeu, Z., Schreckenberg, K., Simons, A.J., Shackleton, S., *et al.* (2007) Trees and markets for agroforestry tree products: targeting poverty reduction and enhanced livelihoods. In: Garrity, D., Okono, A., Grayson, M. and Parrott, S. (eds) *World Agroforestry into the Future*. World Agroforestry Centre, Nairobi, pp. 11–22.

Shackleton, S., Shackleton, C., Wynberg, R., Sullivan, C., Leakey, R., *et al.* (2009) Livelihood trade-offs in the commercialisation of multiple use NTFP: lessons from marula (*Sclerocarya birrea* subsp. *caffra*) in southern Africa. In: Shaanker, R.U., Hiremath, A.J., Joseph, G.C. and Rai, N.D. (eds) *Non-timber Forest Products: Conservation, Management and Policy in the Tropics*. Ashoka Trust for Research in Ecology and Environment, Bangalore, India, pp. 139–173.

Wynberg, R., Cribbins, J., Leakey, R.R.B., Lombard, C., Mander, M., *et al.* (2002) A summary of knowledge on marula (*Sclerocarya birrea* subsp. *caffra*) with emphasis on its importance as a non-timber forest product in South and southern Africa. 2. Commercial use, tenure and policy, domestication, intellectual property rights and benefit-sharing. *Southern African Forestry Journal* 196, 67–77.

Wynberg, R.P., Laird, S.A., Shackleton, S., Mander, M., Shackleton, C., *et al.* (2003) Marula policy brief. Marula commercialisation for sustainable and equitable livelihoods. *Forests, Trees and Livelihoods* 13, 203–215.

Redirecting Agriculture – Going Multifunctional   10

> Tugged at and battered by both sides of this ideological scrap [low-input versus high-energy farming systems], the international agricultural research system is being torn two ways. ... World agricultural science has for two generations been dominated by the high-energy model of farming – and it has produced a food supply miracle, which is now seen to have high social and environmental costs, as well as great dependence on failing resources. Science has largely neglected the equally promising but far less understood low-input systems.
>
> Julian Cribb (2010) *The Coming Famine: the Global Food Crisis and What We Can Do To Avoid It*. University of California Press, Los Angeles, California.

> Food is not the only thing we have to get right – of course not; for everything is connected with everything else. But food is the most pressing issue – day-by-day and even hour-by-hour: the thing we absolutely have to get right. Get farming right, and everything else we want to achieve can begin to fall into place, from the day-to-day pleasures of good eating and social living, to the grand aspirations of full and fulfilled employment, world peace, and the conservation of wildlife. Get agriculture wrong, and everything else is compromised. At present, of all human endeavours, farming is the most ill-fashioned and ill-starred of all, and the ill effects of this are indeed felt everywhere, and by everyone: and the solutions proposed and put in train by the powers-that-be are making things worse.
>
> Colin Tudge (2007) *Feeding People is Easy*. Pari Publishing, Pari, Italy.

As we have seen in early chapters, the need for more sustainable rural development and food production has been on the agenda of the World Summits since 1992, but without any clear idea of how to proceed towards these goals. So, it was with some enthusiasm that I accepted an invitation to contribute to the International Assessment of Agricultural Knowledge, Science and Technology for Development (IAASTD), even though it was unpaid work. The task and challenge posed by IAASTD was to assess how agricultural knowledge, science and technology can be utilized to:

1. Reduce hunger and poverty.
2. Improve rural livelihoods.
3. Facilitate equitable and environmentally, socially and economically sustainable development.

This was to be a large international desk study of how agriculture has contributed to sustainable development worldwide. The process was managed and administered by an international multi-stakeholder bureau based in Washington, DC, ranging in composition from the private sector through research and development organizations, to civil society. The outcomes would be a global report and five regional reports over a period of 3 years. I was asked to be one of two coordinating lead authors responsible for a historical chapter in the global report on the 'Impacts of agricultural knowledge, science and technology on development and sustainability goals'. The authorship of this chapter was a multidisciplinary team that steadily rose to 44 people with widely divergent socio-economic and biophysical expertise and with very diverse work experience.

The principle behind this review was to take account of all viewpoints, to digest all the information down to the bare bones and then to make statements about social, environmental and economic sustainability. The philosophy of IAASTD was not to be prescriptive about what was the best approach, but to look at issues from all sides and to present the options in a transparent and unbiased way. This non-prescriptive approach makes the IAASTD reports very different from all previous reports on agriculture. It was a very difficult process, but one that really forced us to engage with all stakeholders and to develop understanding from many different viewpoints.

Perhaps before going on we need to remind ourselves of some of the points we have seen earlier. The productivity of conventional high-input agriculture has been greatly increased by the achievements of the Green Revolution, saving millions of people from starvation. However, this achievement came at a high environmental cost in terms of land conversion from forest (deforestation), land degradation and the over-exploitation of natural resources – especially soil and water. Agriculture is now also recognized as being a major contributor to climate change. Land clearance, especially on hillsides, also has severe impacts on the hydrological cycle, flooding and soil erosion. Furthermore, despite the success of improved productivity of major food staples, there are still billions of people suffering from poverty, malnutrition and hunger. Sadly and all too frequently, the pressures from population growth have led to situations where the farmers have been forced to cultivate land that has not recovered from previous cycles of shifting cultivation. Consequently they are trapped in food insecurity and poverty and are therefore forced to sow their crops knowing that their soil is spent and that yields will be poor and declining year by year. All of the above have led to many calls for a new approach to food production, especially in the tropics and subtropics where the problems and issues are most urgent and prevalent. The key issues to be addressed are how to achieve land rehabilitation, food and nutritional security and income generation – all within sustainable land use practices.

In our IAASTD chapter, we came up with 290 impact statements that related to sustainable development over the last 50+ years. In this analysis we examined: (i) the achievement of higher yields and better quality crops and farm animals; (ii) ways to protect and even restore natural resources (soil, water, vegetation cover, biodiversity and climate); and (iii) approaches to meeting all the social and economic needs of rural communities. In addition to economics and the more traditional sciences of soil chemistry, crop and livestock physiology, genetics and breeding, we explored the benefits of participatory approaches to problem solving, and the importance of ecological and environmental sciences.

Against the ambitious and exacting criteria of environmental and social sustainability, modern intensive agriculture does not score very highly, although of course intensive high-input agriculture has hugely increased productivity in industrialized countries. However, there are many examples of low-input farming (organic agriculture, as well as conservation agriculture, minimal tillage, ecoagriculture, agroforestry and permaculture), which build natural capital and typically have other positive environmental impacts and so provide pointers towards more sustainable land use practices, especially for the tropics and subtropics.

In addition to the above low-input farming systems, there has been a steady increase over recent decades in the recognition by researchers of integrated approaches to agriculture and rural development (Integrated Pest Management, Integrated Water Resources Management, Integrated Soil and Nutrient Management and Integrated Crop and Livestock Management). These approaches combine ecological principles with more widely recognized areas of agronomy, livestock husbandry and natural resources management. Thus there are some well-defined approaches to developing more socially relevant, pro-poor, smallholder agriculture that: (i) rehabilitate natural capital; (ii) foster ecosystem services; (iii) increase production; and (iv) enhance livelihoods.

Our task therefore was to examine all these different approaches to food production and to score them for their ability to deliver environmental, social and economic sustainability, and to present the evidence in a way that could allow the reader to evaluate the relative merits of different strategies, technologies and practices. To do this we attributed each of our 290 statements with: (i) a degree of certainty (from well-established to speculative); (ii) likely impact (from negative (−5) to positive (+5)) and scale of impact (from local to global); and (iii) any specificities (environmental, geographic or social). These quantifiers were based on the published academic literature.

Not surprisingly, given the diverse backgrounds of the authors and the non-prescriptive nature of our conclusions, the IAASTD reports that emerged from this exercise have a perspective that is very different from the many other reports coming from UN agencies and the World Bank – and indeed very different from the business strategies of the multinational agricultural industries. The global and regional IAASTD reports,[1] which were approved by 61 national governments,[2] are perhaps a more 'grassroots' view of the world.

The IAASTD reports were unique in their recognition that agriculture involves much more than just the production of food. The reports placed great emphasis on the need for agriculture to be seen as multifunctional and to encompass ecosystem services (watershed protection, aquifer recharge, soil fertility, biological diversity, sequestration and emission of greenhouse gases), social services (health, livelihoods and cultural practices) and economic services (marketing, trade, employment, etc.) and not just crop cultivation, livestock production, forestry and fishery that produce food, feed, fibre, fuel and other goods.

In broad terms, multifunctional agriculture recognizes the 'inescapable interconnectedness of agriculture's different roles and functions' in rural development. Consequently, multifunctional agriculture is the means to make significant progress towards the integration of the highly complex set of factors conferring sustainable rural development. Thus multifunctional agriculture is aimed at simultaneously targeting the alleviation of hunger with the improvement of livelihoods through improved health and nutrition, economic growth, and enhanced social and environmental sustainability. Multifunctional agriculture is therefore strongly dependent on diversification, both in terms of the agroecosystem and in terms of the range of products contributing to income generation in a diversified rural economy.

One consequence of the uniqueness of the IAASTD reports is that they have been criticized and condemned by some with very different views – most vociferously by big business and some branches of academia. Specifically, we were accused of being negative, anti-science and anti-commerce. The charge of negativity is not borne out by analysis of our impact scores, which illustrate that our review spanned both positive and negative outcomes and that overall our findings were tilted towards the positive. The other charges seem to have been the result of presenting all sides of the biotechnology debate rather than just accepting genetically modified organisms as a silver bullet that will solve the food crisis.

It is appropriate to point out that the IAASTD reports followed a series of other internationally endorsed assessments that have alerted the world to the role of agriculture in climate change (International Panel on Climate Change), ecosystem and environmental degradation (Millennium Ecosystem Assessment), water scarcity (Comparative Assessment of Water Management in Agriculture) and natural resource degradation (Global Environment Outlook). All of these reports have singled out agricultural practices as being responsible for much of the damage to natural resources and the global environment and the continuing severity of poverty, malnutrition and poor human health. The IAASTD reports also strongly endorsed the capacity of good agricultural practices to restore natural resources, enrich agroecosystems and rehabilitate degraded environments in ways that also benefit mankind. Thus, many of the outcomes of IAASTD are congruent with the Millennium Development Goals and the Convention on Biological Diversity and are compatible with other international initiatives, like the World Summit on Sustainable Development's Water, Energy, Health, Agriculture and Biodiversity initiative.

In his recent book, *The Coming Famine*, Julian Cribb[3] makes the very relevant point that global agriculture has to embrace both sides of what he describes as the philosophical divide between adherents of 'low-input smallholder agriculture' and 'high-energy farming systems'. He specifically recognizes that each of these approaches to agriculture is appropriate in different places and circumstances. This book – *Living with the Trees of Life* – mainly addresses the former because of the special relevance of low-input farming systems like agroforestry to smallholder farmers in the tropics. This is because of the current socio-economic and environmental constraints to tropical agriculture that we have examined in earlier chapters. However, I fully acknowledge that once these constraints are overcome and the very poor are on a path out of poverty, the responsible use of fertilizers, other agrichemicals and genetically modified crop varieties may become increasingly available to farmers and appropriate for tropical farming systems. Likewise, to reduce some of the negative impacts of high-energy farming, low-input systems may also become more relevant to agriculture in industrialized countries.

The over-riding questions that need to be addressed are:

- How can the land be used to feed a growing population without further damage to the local and global environment?
- How can food and nutritional security be improved on a declining area of available land?
- How can the land be used to enhance the livelihoods and income of those in poverty?

To see a way forward, let's look at the issues in more detail. First, and critically, the world is running out of good fertile land for agriculture without cutting down what remains of our forests. Further deforestation is an unacceptable option, so we have to bring degraded land back into production.

Secondly, current land use practices in the tropics have led to deforestation, overgrazing and over-exploitation of soils and water resources (Fig. 2.1), causing a cascade of negative impacts: land degradation, loss of soil fertility, loss of biodiversity, the breakdown of agroecosystem function, and declining yields – creating the so-called yield gap (Fig. 10.1a). Together these things result in increased hunger, malnutrition and declining livelihoods. Associated with this is reduced access to traditional wild foods, loss of income, and the increased need for costly (often unaffordable) agricultural inputs. In addition, the widespread clearance of forest from the landscape, especially from hillsides, exposes soils to erosion and increases runoff, resulting in landslides and flooding that destroy property and cause the death of large numbers of people. Loss of perennial vegetation also contributes to climate change.

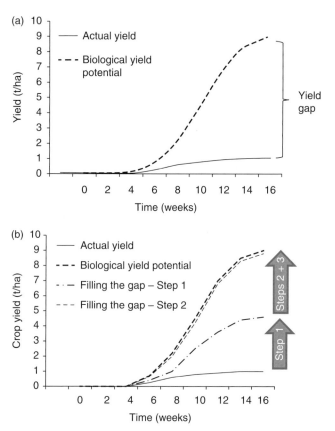

Fig. 10.1. (a) The yield gap is the difference between the biological potential of the crop variety and the yield actually attained by farmers. (b) Steps to filling the yield gap (Step 1 = biological nitrogen fixation from fertilizer trees; Steps 2 and 3 = domestication and commercialization of indigenous fruit and nut trees).

So, to find a way forward I believe we need to recognize four important points:

1. Many farmers in the tropics are failing to grow staple foods anywhere near their existing biological potential, creating the yield gap – the difference between the potential yield of a crop and the yield actually achieved by farmers.
2. Poor smallholder farmers living in poverty cannot afford to buy the fertilizers and pesticides (even if they had adequate access to them) that would allow them to practise monoculture agriculture.
3. The overriding dominance of starchy food staples in modern agriculture may provide adequate calories for survival, but they lack the protein and micronutrients for healthy living, not to mention the sensory pleasures of the traditional and highly nutritious foods that used to be gathered from the forest.
4. Many food crops have more or less reached their peak biological yield potential already.

So, how do we find our way around these problems? Clearly a focus on crop yield is important, but rather than trying to increase yield potential, let's concentrate on the big problem of the yield gap. The primary cause of this gap is poor crop husbandry, which leads to the loss of soil fertility and agroecosystem functions, such as: (i) the cycling of nutrients, carbon and water; (ii) the progress and fulfilment of life cycles and food webs that maintain the natural balance between organisms; and (iii) the dispersal of pollen and seeds for effective reproduction.

In the worst cases of land degradation, maize farmers are achieving yields of only 0.5–1.0 t/ha when the potential is around 10 t/ha – a 9 t gap. In this situation, closing the gap could increase food production by ten- to 20-fold. However, if even only a two- to threefold increase was achieved on average, this is still well over the 70% increase that will be required to feed the 9 billion people predicted[4] to populate the world by 2050. Is this attainable?

Typically, soil nitrogen is the prime constraint to crop growth and high yields on degraded soils. In some soil types, severe nutrient impoverishment cannot be restored by fertilizers until the soil organic matter has been enriched by manures, mulching and fallows. Generally, however, the levels of soil nitrogen can be restored by harnessing the capacity of certain legumes to fix atmospheric nitrogen in root nodules colonized by symbiotic bacteria (*Rhizobium* spp.). As we saw in Chapter 3, numerous techniques have been developed to integrate appropriate legume species within farming systems. Probably the most effective and adoptable are high-density improved fallows with species like *Sesbania sesban* and *Tephrosia vogelii* or relay cropping with *Gliricidia sepium*. Leguminous crops like beans and groundnuts can also contribute to this process. Together the legumes will increase soil nitrogen to a level that will give maize yields of 4–5 t/ha within 2–3 years. I should make it clear, however, that the capacity of these leguminous plants to improve soil nitrogen is not the same in all soil types. However, in general their use can lead to the partial closure of the yield gap and so to increased food security.

It is perhaps worth reflecting that temperate agriculture took a somewhat similar step when in the early 1800s, after the Battle of Waterloo, the three-field crop rotation was supplemented with a fourth field, a leguminous fodder crop of clover with ryegrass, which was then called the Norfolk System in the UK. In addition to restoring soil fertility, this had the added benefit of giving a boost to livestock farming.

The use of legumes to restore soil fertility also starts the process of restoring agroecological function. For example, one of the serious weeds of cereal crops like maize, millet and sorghum is *Striga hermonthica*. It is a root parasite on these cereals and its seeds germinate in response to root exudates from the young cereal plants. Interestingly, however, *S. sesban* and the fodder legumes *Desmodium intortum* and *Desmodium uncinatum* also trigger *Striga* germination, so they can be used to promote suicide germination in the absence of the cereal hosts. *Desmodium* spp. also act as a repellent to insect pests of cereals, for example the stem borers *Busseola fusca* and *Chilo partellus*.[5] Likewise simple agroecological benefits can also be attained by planting napier grass (*Pennisetum purpureum*) as an intercrop, or around small fields, as it attracts the pests away from the crops.

Going to a fully functional and more diverse agroecosystem involves the integration of trees within the farming systems. Some trees are of course cash crops like coffee, cocoa and rubber, which in the past were either grown as large-scale monocultural plantations or as a two-species mixture, such as cocoa under the shade of coconuts or *G. sepium*. Increasingly, however, they are becoming smallholder crops grown in much more diverse species mixtures, such as bananas with fruits trees like mango, avocado and local indigenous trees producing marketable products. This is well developed in Latin America and is becoming widely recognized as a way to restore the biodiversity normally found in natural forests. Likewise in West Africa we saw that the replacement of shade trees with trees that also produce useful and marketable products is a good strategy for farmers wanting to maximize output from the land and to minimize the risks associated with reliance on a single crop species.

Furthermore, we also saw in Chapter 3 that the ultimate example of this diversification process has occurred in a farmer-led, silent revolution in Southeast Asia by farmers replacing the natural forest fallows of shifting cultivation with commercial tree fallows (agroforests – Figs 3.6 and 3.7 illustrate diversified agroecosystems at maturity) on the valley slopes. This creates a land use mosaic such as that found in temperate landscapes. Using this approach to agriculture, food crops are grown on the best and most fertile land, and the more marginal land is used for income generation from tree crops (Fig. 10.2). These tree crops also protect sloping land from erosion, improve

Fig. 10.2. A multifunctional agricultural landscape in Vietnam.

water infiltration into the soil, sequester carbon so mitigating climate change, enhance biodiversity and promote agroecosystem function. Additionally, as mentioned in Chapter 4, there is also some evidence that complex perennial vegetation is better than an herbaceous crop at recycling moisture to the atmosphere to be advected downwind to fall as rain. It therefore seems likely that agroforests in the wet tropics would be beneficial to rainfed agriculture in dry and drought prone areas further downwind in the continental interior.

In summary, therefore, soil fertility amelioration with nitrogen-fixing leguminous plants and diversification of the farming system with tree crops are two interventions that can be used to improve food security by restoring soil fertility and initiating an agroecological succession that rehabilitates farm land and reverses some of the land degradation processes. We can think of this as the first step towards closing the yield gap (Fig. 10.1b). The complete closure of the yield gap, however, typically requires the provision of inorganic nutrients, as other soil nutrients, such as phosphates, are often limiting. These fertilizers have to be purchased. So, the need now is to generate income. One source of income for farmers is payments for environmental services – things like carbon storage in trees, watershed protection and biodiversity conservation by forests and agroforests. However, the products from trees themselves can also be a source of income.

The good news here is that by diversifying the farming systems with commercially important trees, we have already initiated the second and third steps towards closing the yield gap by domesticating and cultivating a range of indigenous trees for their marketable products – products from trees can be harvested and sold to enhance household livelihoods. This, however, is just the beginning, as the completion of these steps leads to the intensification of multifunctional agriculture by creating new high quality tree crops. We have already seen in Chapters 5–8 how this can be quickly and easily achieved to improve uniformity and market demand. These domesticated trees can then be used to enrich and improve the farming systems, whether as shade for commodity crops, agroforests on hillsides, orchards, field and farm boundaries, fodder banks, or woodlots. The range of options of species and location is infinitely variable and will differ between farms across a landscape. There is, therefore, no prescription about how to implement this approach to agriculture. It is up to the farmers and their communities to do what suits their particular situation. This is an infinitely adaptable model.

We should not forget the importance of livestock in agriculture. The 2020 projections of the International Food Policy Research Institute suggest that we will need 40% more grain and will eat a lot more meat. As we have just seen we can greatly increase grain production by closing the yield gap. We have also seen that there are opportunities to use fodder trees to increase the productivity of cattle and goats. The integration of fodder trees and livestock into a farm is one of the elements of diversification that could be part of Step 2. Ideally, to maximize the benefits of multifunctional agriculture both crop and livestock diversification should occur together in a truly mixed farming system that both enhances self-sufficiency and creates economic growth.

In multifunctional agriculture, intensification comes with enhancing the productivity of the diverse set of species filling all the various ecological niches in the agroecosystem. In other words, we are talking about 'polycultural intensification'. Put this way we have a lot to learn about how to intensify. We had a glimpse of the challenge that this poses to modern science when we discussed the diversification of cocoa agroecosystems in Chapter 4. There is also a challenge in this way of thinking for economists who focus avidly on the adage of 'economies of scale', which maximize the benefits of investment for the individual, a company or even a nation. For poor smallholder farmers, risk aversion through diversification is a much more appropriate production strategy.

The income from these trees can obviously be spent in many different ways depending on the household priorities – for example a mobile phone, new cow or motor bike, children's health or education, family wedding or funeral, or numerous other things. However, one option is to spend it on fertilizers and agricultural inputs and so to close the yield gap by raising cereal yields up towards their biological potential of around 8–10 t/ha.

This further increase in food crop yield may once again increase the area available to new cash crops for further income generation.

To summarize this approach to addressing the cycle of land degradation and social deprivation (Fig. 2.1), we can enhance food security by using leguminous trees and shrubs to enrich soil nitrogen and initiate agroecological function and then diversify the cropping system with new tree crops to generate income from the sale of their products and to further improve agroecological functions. The new tree crops can in some circumstances also generate additional income from payments for environmental services. Finally, we can commercialize the products from the new crops to promote local business and employment. So, by simultaneously addressing the two major driving forces behind the loss of crop yield – poverty and land degradation – we can put all this together as a package that should go a long way towards slowing down or reversing the cycle of land degradation and social deprivation. Through the inclusion of traditionally important fruits and nuts, this approach also provides invaluable micronutrients in short supply in our staple food crops. Thus if scaled up and out, this approach to closing the yield gap should make a very substantial contribution to alleviating poverty, hunger and malnutrition and also reduce the negative impacts of agriculture on the environment.

In contrast to what I have been presenting, there are proponents of redoubling our efforts on intensive, high-input industrial farming to further increase the yield of staple food crops by enhancing their capacity to withstand biotic and abiotic stress. Breakthroughs in molecular plant genetics, such as genetic modification, may indeed find new and exciting ways to improve crop yields. For example, in my maize-crop scenario, if ways could be found to add the capacity to fix nitrogen to maize plants, this would go some way towards achieving Step 1 of filling the yield gap. This would be further enhanced if these genetically modified maize varieties were also more resistant to pests like cereal stem borers and the parasitic weed *Striga*. As we have seen, achieving Step 1 of closing the yield gap would greatly improve both local and global food security and would be a huge leap forwards for agricultural production. However, it still leaves the need to complete Steps 2 and 3, which are all about income generation and growth in the rural economy for employment and business development. This is needed for farmers to be able to afford to purchase, among other things, agricultural inputs. I believe that agroforestry and the domestication of the Trees of Life remain a highly appropriate and desirable way to stimulate sustainable economic growth in tropical countries.

It is clear that we will need to use the full arsenal of agricultural technologies based on the full spectrum of first-rate science available to us, if we are to feed and support the livelihoods of a global population of 10–12 billion in the future. It is my view that the widespread adoption of multifunctional agriculture should involve the big agribusiness companies marketing fertilizers, pesticides and seeds. Indeed, I think there would be great benefits for them too if they were to promote this much more sustainable approach to agriculture. It seems to me that multifunctional agriculture would create a new generation of farmers entering the cash economy – potentially billions of them! They would become potential customers. To date, the agribusiness companies do not appear to have seen this possible way forward. However, if these powerful companies were to recognize this potential and get behind a multifunctional approach to scaling up the resolution of the land degradation issues, the chances of having meaningful impact on the big social, economic and environmental issues would be greatly increased. Let's be clear: agroforestry reduces the need for many agrichemicals but, unlike organic agriculture, has always seen a role for the wise use of agrichemicals. Agroforestry does, however, promote the biological alternatives because they are: (i) affordable and socially more relevant to poor farmers; (ii) better for the environment; and (iii) central to the rehabilitation of degraded farmland and to watershed protection.

There is another advantage of going the multifunctional route to agricultural reform, a political advantage that should make accountants and development agencies happy. If farmland can be made more productive so that crops give good yields close to their biological potential, then we will be able to reap greatly improved returns from the huge international investment in crop breeding over the last 40–50 years under the Green Revolution. Getting better returns on previous investments by better crop husbandry and closing the yield gap should be welcomed by donors. If the yield gap was closed in this way on a scale that raised 1, 2 or, better still, 3 billion farmers out of food insecurity and poverty, the world's food crisis would be averted and there might even be an embarrassing surplus!

The production of these marketable products is the start of a value chain that creates wealth and economic growth through product processing, packaging, value adding and trade. This is where the greatest opportunity to alleviate poverty kicks in. Through the creation of local business opportunities and employment down the value chain, the number of people engaged in agricultural production can be reduced. Currently in Africa, some 80% of the rural population is engaged in farming, while in Europe the equivalent proportion is about 1–3%, with many of the remaining 97–99% engaged in other forms of employment, creating wealth for the household and the overall population. In the tropics we still have a very long way to go. However, we will see in the next chapter that the relatively recent initiatives to develop new cash crops based on the domestication of highly nutritious indigenous foods for local and regional markets are moving us in the right direction. Likewise, we will also see in the next chapter that commercial activities and value adding are expanding the marketing opportunities in Cameroon for tree products, and that this is transforming the lives of the participating communities.

In Chapter 9 we saw that there are a number of issues around the commercialization of products from newly domesticated tree crops that need to be resolved to ensure that the farmers are the beneficiaries of their hard work. Most of the traditionally important products from tropical forests have been marketed locally for centuries. Over the last decade an increasing number of the traditionally used and marketed tree products have entered national, regional and international trade as new processed food, medicinal, nutraceutical and cosmetic products, based on the fruits, nuts, gums, resins and fibres (Fig. 9.1), creating market demand and economic growth.

The marketing and trade of commodities from tropical producers have often been exploitative, so the emergence of initiatives from organizations like the Fairtrade Foundation and PhytoTrade Africa to ensure that the producers receive a fair price through long-term trade agreements is encouraging. Another encouraging development has been the emergence of PPPs between local producers in developing countries and far-sighted, multinational companies. These partnerships typically have a strong development focus, working to invest and develop industries to promote local value adding, marketing and trade actually in developing countries.

A large part of the problem in agriculture today is the much bigger problem of the consumptive philosophy of modern society, which extends to economic development nationally and internationally. The fundamental principle of drawing-down stocks of natural capital has got us into the mess we are in. Can we find the political will to fight against this philosophy?

In conclusion, agriculture is truly at a crossroads. We can doggedly go on in 'business as usual' mode or we can change direction towards multifunctional agriculture, delivered by agroforestry (Box 10.1). I hope this chapter points the way to closing the yield gap and averting further series of food crises that leave huge numbers of people hungry, malnourished and destitute. To change direction, I believe that the developing countries will have to take the lead and call for the acceptance of multifunctional agriculture, as I doubt it will happen if the decision making is left to the developed countries. Indeed, all the signs from policy debate[6] are that the options being considered to overcome the current international 'food crisis' are set to favour the rich countries – again – and not the poor.

Some rich countries have a different agenda, despite all the rhetoric about the need to reduce poverty, malnutrition, hunger and disease. Developing countries are going to have to make their voices heard in Washington, New York, London and Brussels if they want a different outcome from the current round of policy discussions. I hope this book provides some inspiration for a more multidisciplinary approach to agriculture and rural development.

> **Box 10.1.** The characteristics of agroforestry important for the delivery of multifunctional agriculture.
>
> The features that agroforestry brings to multifunctional agriculture to enhance social, economic and environmental resilience in agriculture and rural development are:
>
> - based on traditional knowledge and culture;
> - based on participatory techniques to ensure relevance to local people;
> - based on integrated natural resources management and sustainable land use; and
> - based on knowledge of the natural resource.
>
> The above features mean that multifunctional agriculture can:
>
> - empower subsistence farmers to control their destiny;
> - enhance food security and rural/urban livelihoods, reducing hunger;
> - enhance nutrition security and health, reducing malnutrition and diseases;
> - enhance opportunity for income generation, reducing poverty;
> - diversify farming system at the local and landscape scale, enhancing watershed services and sustainable production;
> - create new agricultural commodities;
> - diversify market economy and buffer commodity price fluctuations;
> - decentralize business opportunities to the villagers;
> - create employment in processing and marketing;
> - build social responsibility from the 'grassroots';
> - enhance international public goods and services, reducing climate change and loss of biodiversity;
> - offer opportunities for new policy interventions to combat deforestation, desertification and land degradation; and
> - break down the disconnects between disciplines and organizations responsible for policy and its implementation in rural development.

Notes

[1] Edited by Drs Beverley McIntyre, Hans Herren, Judi Wakhungu and Bob Watson.
[2] Three countries (the USA, Canada and Australia) recorded a small number of reservations about specific sections of the text.
[3] Julian Cribb (2010) *The Coming Famine: the Global Food Crisis and What We Can Do To Avoid It*. University of California Press, Los Angeles, California.
[4] International Food Policy Research Institute (2011) Press release 24 May 2011 by Mark Rosegrant at Ag Innovation Showcase, St Louis, Missouri.
[5] This has been called the push–pull technology – Khan *et al.* (2000) *Pest Management Science* 56, 957–962.
[6] Julian Cribb (2010) *The Coming Famine: the Global Food Crisis and What We Can Do To Avoid It*. University of California Press, Los Angeles, California.

Further Reading

Beintema, N., Bossio, D., Dreyfus, F., Fernandez, M., Gurib-Fakim, A., *et al.* (2008) In: McIntyre, B.D., Herren, H., Wakhungu, J. and Watson, R.T. (eds) *Global Summary for Decision Makers*. International Assessment of Agricultural Science and Technology for Development (IAASTD). Island Press, New York.

Butler, L.M., Leakey, R.R.B., Albergel, J. and Robinson, E. (2008) Natural resources management. In: McIntyre, B.D., Herren, H., Wakhungu, J. and Watson, R.T. (eds) *IAASTD Synthesis Report*. International Assessment of Agricultural Science and Technology for Development (IAASTD). Island Press, New York, pp. 59–64.

Gurib-Fakim, A., Smith, L., Acikgoz, N., Avato, P., Bossio, D., *et al.* (2008) Options to enhance the impact of AKST on development and sustainability goals (Chapter 6). In: McIntyre, B.D., Herren, H., Wakhungu, J. and Watson, R.T. (eds) *International Assessment of Agricultural Science and Technology for Development: Global Report*. Island Press, New York, pp. 377–440.

Kiers, E.T., Leakey, R.R.B., Izac, A.-M., Heinemann, J.A., Rosenthal, E., *et al.* (2008) Agriculture at a crossroads. *Science* 320, 320–321.

Leakey, R.R.B. (2001a) Win:win landuse strategies for Africa: 1. Building on experience with agroforests in Asia and Latin America. *International Forestry Review* 3, 1–10.

Leakey, R.R.B. (2001b) Win:win landuse strategies for Africa: 2. Capturing economic and environmental benefits with multistrata agroforests. *International Forestry Review* 3, 11–18.

Leakey, R.R.B. (2003) The domestication of indigenous trees as the basis of a strategy for sustainable land use. In: Lemons, J., Victor, R. and Schaffer, D. (eds) *Conserving Biodiversity in Arid Regions*. Kluwer Academic Publishers, Boston, Massachusetts, pp. 27–40.

Leakey, R.R.B. (2010) Agroforestry: a delivery mechanism for multi-functional agriculture. In: Kellimore, L.R. (ed.) *Handbook on Agroforestry: Management Practices and Environmental Impact*. Environmental Science, Engineering and Technology Series. Nova Science Publishers, New York, pp. 461–471.

Leakey, R.R.B. (in press) Smallholder cocoa agroforests: multiple forest products and services – a model for sustainable development? In: Bennett, A.B., Keen, C., Shapiro, H. and Schroth, G. (eds) *Theobroma cacao: Biology, Chemistry and Human Health*. Wiley-Blackwell, Oxford.

Leakey, R.R.B., Kranjac-Berisavljevic, G., Caron, P., Craufurd, P., Martin, A., *et al.* (2008) Impacts of AKST on development and sustainability goals. In: McIntyre, B.D., Herren, H., Wakhungu, J. and Watson, R.T. (eds) *International Assessment of Agricultural Science and Technology for Development: Global Report*. Island Press, New York, pp. 145–253.

Lombard, C. and Leakey, R.R.B. (2010) Protecting the rights of farmers and communities while securing long term market access for producers of non-timber forest products: experience in southern Africa. *Forests, Trees and Livelihoods* 19, 235–249.

Schreckenberg, K., Barrance, A., Degrande, A., Gordon, J., Leakey, R., *et al.* (2005) Trade-offs between management costs and research benefits: lessons from the forest and the farm. In: Holland, J. and Campbell, J. (eds) *Methods, Knowledge and Power: Combining Quantitative and Qualitative Development Research*. ITDG Publishing, London, pp. 191–204.

Multifunctional Agriculture – Proof of Concept

11

> Put simply, participatory tree domestication refers to the means by which rural communities select, propagate and manage trees according to their own needs, in partnership with scientists, civic authorities and commercial companies. It is usually oriented towards specific local markets and encompasses the use of both indigenous knowledge and genetic selection based on scientific principles.
>
> Zac Tchoundjeu *et al.* (2006) Putting participatory domestication into practice. *Forests, Trees and Livelihoods* 16, 53–69.

> Given the long lag time between investment in agricultural research and the resulting production increases, failure to invest today will show up in production shortfalls and environmental degradation 10 to 20 years from now. The problems associated with environmental degradation will present themselves sooner. We must not wait until a global food crisis is upon us, or until the last tree has fallen before making these investments.
>
> Per Pinstrup-Andersen (1994) *World Food Trends and Future Food Security.* International Food Policy Institute, Washington, DC.

In some quarters the holistic approach to multifunctional agriculture is seen to run counter to intensification and to be turning the clock back to a lower level of productivity. This is a misunderstanding. My response to the advocates of this argument is that we need a different kind of agricultural intensification, one that meets the needs of people for better nutrition and health, enhances livelihoods from economic diversification into new areas of employment and business, and brings degraded land back into production. So, as we have seen earlier, the problem is much more complex than just producing more staple foods, as it has critical social, economic and environmental dimensions. It also requires a change in mindset away from the over-exploitation of natural resources – especially soil and water. On top of all this people want to sustain their traditions and culture.

Intensification is often portrayed as synonymous with simplifications such as monoculture. This book has presented an alternative intensification strategy, one based on diversification into a new set of crops to be grown along with the conventional staple food crops, in ways that allow them to express their, often hidden, genetic potential. We have also seen that tree domestication leads to intensification in mixed farming systems through improved yields and quality for increased income generation and for new industries that create opportunities for value adding, business and trade, especially for the poor and needy of the tropics and subtropics.

The concepts and practices that I have described in this book are based on over 25 years of research by the World Agroforestry Centre, its partners and a growing number of other agroforestry research teams in academic and development organizations around the world. In this chapter we look at a project that has been designed to pull together three key elements of agroforestry and multifunctional agriculture: (i) restoration of soil fertility by nitrogen fixation; (ii) tree domestication; and (iii) the commercialization of tree products. This case study – the Food for Progress project – is an integrated rural development programme in the North and North-west regions of Cameroon. I believe it is the best example to date of 'proof of concept', although there are many other good examples of agroforestry improving the lives of smallholder farmers, some of which were mentioned in Chapter 3.

The Food for Progress project grew out of the tree domestication research in Cameroon, and especially the development of the concept of participatory tree domestication. I must reiterate and emphasize again that this project is not a prescription of how to implement multifunctional agriculture using agroforestry. It is basically an example of one of an infinite number of combinations of species and agroforestry practices that can be developed to meet the requirements of different social, economic and environmental situations found around the world. It therefore illustrates the model presented in *Living with the Trees of Life*.

'Food for Progress' was implemented as a development project shortly after I left ICRAF. It was masterminded by Dr Zac Tchoundjeu and his team, who provided assistance in the form of training and mentoring to a group of NGOs that were responsible for project implementation. So this was the start of putting the theory and the research outputs into practice. The project was started with funds from the International Fund for Agricultural Development (IFAD) on a small scale in five pilot villages in 1998. These first nurseries played the role of central or training nurseries for local farmers. Their role has since been expanded and they have become Rural Resource Centres (RRCs), enhancing the capacity of community members to engage in agroforestry by providing training on topics such as: (i) the use of 'fertilizer trees' to restore soil fertility; (ii) tree propagation and nursery management; and (iii) tree domestication using simple low-technology horticultural techniques.

Through this training programme the RRCs have built a strong foundation in the communities that has empowered the farmers to become self-sufficient. With time, these RRCs have developed their own 'satellite' nurseries around the region (Fig. 11.1). These satellite nurseries have been developed as farmers from neighbouring villages have acquired the necessary skills to create their own facilities.

The successful expansion of the overall programme has resulted from conventional techniques of dissemination, such as the media and posters, as well as more innovative approaches like the organization of exchange visits between communities. The latter allows those thinking about becoming involved to go and see for themselves what other farmers have already achieved, as well as giving them the opportunity to discuss

Fig. 11.1. A village nursery in Batibo, Cameroon – one of the satellite nurseries of the Food for Progress project.

the pros and cons with those who have already engaged in the process. Within Cameroon, some of these exchange visits were further promoted by competitions between nurseries covered by the radio and TV.

Ten years after the start of the project, some new components were added when financial support changed to the United States Department of Agriculture (USDA) in partnership with the Cameroon Ministry of Economy, Planning and Regional Development (MINEPAT). These new components included partners to promote: (i) community project management; (ii) infrastructure development; and (iii) marketing and business skills, including the use of microfinance and the fabrication of processing equipment for agricultural produce.[1] At first sight this sounds like a somewhat curious combination, but when they become integrated into a self-help package for poor smallholder farmers it becomes clear that in reality they are highly compatible. Indeed, these components are synergistic in their interactions and create the stimulus to be entrepreneurial and add value to both conventional agricultural produce, such as cassava, and to new agroforestry tree products obtained from indigenous trees that have been ignored by science.

Microfinance allowed the villagers to expand or intensify their on-farm activities, while the processing equipment created new off-farm business and employment opportunities in value adding and marketing. The community development training stimulated community-level infrastructure developments such as piped water supplies, wells, roads, bridges and storerooms. These three things stimulated innovation in the communities.

The driving force behind the communities' enthusiastic participation was still the chance to improve livelihoods through: (i) increased farm production; (ii) better nutrition and health; and (iii) the generation of income from domesticated tree products. Together this exposure to new opportunities opened up incentives for further economic growth in the community. The life-transforming potential of this combination of synergistic factors has been motivating.

The concept behind the Food for Progress project was that progress towards sustainable rural development could be stimulated by multifunctional agriculture through the implementation of the three agroforestry steps to closing the yield gap (Fig. 10.1b). This represents an infinitely adaptable generic model for agroforestry. It is important to realize that through agroforestry, each of these steps in the generic model can be implemented in diverse ways, depending on the different species, environments and social and economic situations that prevail at any particular site. We will see that by implementing these three steps, poor rural people really do have the capacity, through the diversification and intensification of agroforestry, to improve their livelihoods and to begin the climb out of poverty, malnutrition and hunger. However, to be successful on any meaningful scale will require a massive programme of capacity building to bring the relatively simple suite of agroforestry technologies to billions of poor farmers. Essentially, this is what we see starting to happen in the integrated rural development Food for Progress project of Cameroon.

When I visited the project in 2009, I had the opportunity to see how this small seed of an idea had grown into a successful integrated rural development package. There were now more than 485 villages, involving over 7000 farmers associated with five RRCs and their 123 satellite tree nurseries in surrounding communities. As the number of participants expanded, they were supported by NGOs and community-based organizations that provided agroforestry and tree domestication training and mentoring at the village level. On my tour around the region, I found engaged and enthusiastic farmers who had become adopters of participatory domestication and, even more importantly, they were starting to make money from the sale of plants and improved cultivars from their nurseries. Some farmers were already planting their fruit trees. With up to 120 trees established on their individual farms, it is likely that production will very soon exceed domestic needs and so the farmers will be able to further increase their income by selling fruits from their cultivars. To enhance interest in and ownership of these cultivars, the best will be named in ways that recognize the farmer and the community that produced them. However, to date, fruit sales have not been the priority and most of the trees planted so far have been for domestic consumption.

Experience has shown that in the first 2 years the output from a village nursery is mainly planted within the community to satisfy the needs of the farmers themselves. Thereafter, however, some of the output can be sold to neighbours and to others outside the village, starting the flow of much-needed income. At first the sales have mostly been of 'fertilizer' trees for soil fertility enhancement. For example 80% of the plants produced were leguminous trees and shrubs as improved fallows are a well-accepted technology in most of the communities engaged in this project. Using these techniques, farmers have reported that their crop yields had doubled or trebled. This is a significant increase in the productivity of the staple food crops and does much to increase food security. One additional benefit from this activity is that many leguminous trees and shrubs are also popular with bees and so many communities have also taken up beekeeping. As a result, in some communities everyone now has access to honey.

The other 20% of the plants being produced are indigenous trees producing fruits, nuts and medicines selected on the basis of the farmers' local knowledge. At this point, the farmers have three choices with regards to how to proceed. They can either put all their efforts into multiplying their plants by vegetative propagation, or they can plant them in their own farms so that they will soon have marketable fruits. Alternatively, they can become a commercial nursery selling plants to other farmers. The choice made will affect the speed and scale of the income generation. The first of these options gives the least 'returns' in the short term, but by far the largest 'returns' in the longer term, as the maximization of returns is in effect a numbers game and the propagating option is the route to the greatest numbers. Usually, however, the farmers opt for splitting their propagated plants among the three options, so ensuring that they build up numbers for the future as well as starting to produce their own fruits and nuts for home consumption. This option also means that some plants can be sold to generate instant income. Evidence shows that the volume and value of these sales is steadily climbing year by year (US$145, US$16,000 and US$28,350 on average after 2, 5 and 10 years, respectively), but the numbers sold vary between the different RRCs and nurseries.

When putting the trees back into the landscape, no two farmers will do exactly the same thing, as both they and their farms will differ from their neighbours. Consequently, for their given circumstances, each farmer will think about where to put their trees – some will scatter them across the farm, others will clump them in corners of marginal land, or spread them out in lines along the contours or around the field boundaries. Some will put them in their homegardens, where they cultivate all sorts of useful plants for everyday use as foods, medicines and other products. Some may just put one tree outside the back door! When you add to this the wide range of potential species that can be included in the farming system, it becomes very obvious that diversity is the order of the day, which is good for risk-averse, resilient farming systems and the creation of agricultural landscapes in which trees are a dominant feature (e.g. Fig. 10.2).

At Njinikejem, near Belo, the Twantoh Mixed Farming Common Initiative Group was one of the original pilot villages 12 years ago. This group of farmers successfully combined improved fallow technologies with the tree domestication techniques. Their nursery has become a small enterprise with demonstration plots containing different high-value fruit trees of the region. In these demonstration plots, cuttings of *Prunus africana* grew rapidly and marcotts of *Dacryodes edulis* and *Cola nitida* fruited 2–3 years after planting. This early fruiting provided a great incentive to farmers who realized that they could easily generate substantial income from their nursery activities. This was illustrated best at Koung-Khi, where the income from the PROAGRO Resource Centre nursery was about US$40,000 in 2009. From these examples, it was clear that in the early days of the development project, planting stock derived as cultivars from superior trees was the biggest source of income in the satellite nurseries. This capacity to generate income from nurseries can be developed quite rapidly. For example, at Batibo a nursery set up under the Promotion of Woman's Initiative in Self-Help Development, which had existed for only 8 months at the time of my visit, was already full of marketable plants of 15 species.

Other income-generating activities supported under the Food for Progress project include several cassava-processing mills, which are operated by groups of women. The largest of these groups was run by ten women who employed eight workers and it processed about 66 bags of dried cassava flour per day, each bag weighing 180 kg. Gabonese traders were buying these bags at US$40–54 per bag, depending on the season. Profits were estimated to be US$2.70 per bag, so as this unit was working throughout the year it suggests that each of the ten women was making profits of around US$3000–4000 per year.

One of the constraints to better food processing being tackled by the project is the availability of local machinery. So through the involvement of Winrock International, several local metal workers have been helped to develop appropriate equipment for drying, chopping and grinding a range of foodstuffs, including spices and some new agroforestry products not previously processed. The metal workers had benefited from the provision of improved designs and from the sales of their machines in local towns (10–20% profits), while local entrepreneurs and producers were benefiting from the use of this equipment to extend the shelf life and quality of their produce. For example, one entrepreneur had set up a stall in Bamenda market (Fig. 11.2) selling sealed packages of high quality dried herbs made from indigenous plants, mostly agroforestry trees (njangsang – *Ricinodendron heudelotii*; bitter leaf – *Vernonia* spp.; eru – *Gnetum africanum*). He told us that he sold 150 g bags of eru for

Fig. 11.2. An innovative stallholder in Bamenda market, Cameroon, who processes and markets products from indigenous trees.

US$1.35. Although this is a considerable 'mark-up' on fresh eru, his trade had increased threefold in 4 months, and the traditional traders in the market were becoming jealous. This small business had also created a few new jobs. Benefits like this are not automatic. For example, when we visited another entrepreneur who was processing chilli peppers, garlic and ginger, we saw he had a stockpile of unsold products and heard how he needed new market outlets. Without these markets, the income generation opportunities are limited. Sometimes specialist help is needed to identify market opportunities, as supply chain development takes time and effort, as well as special skills.

We can see therefore that both the fabrication workshops for the small tools and equipment and the value-adding activities based on product processing have created new employment opportunities outside agriculture, which are spawning local industries to diversify and enhance the rural economy, with numerous knock-on benefits for the population.

One other important activity instigated by the project to improve farmers' financial situation is the provision of short-term and small-scale loans for the purchase of inputs such as seeds, fertilizers and hired labour. In the first phase of loans, US$78,000 was made available to over 900 farmers, 70% of whom were women, in 82 communities. While there were some issues about the mismatch between the terms and conditions of these loans, it was very evident that the farmers had benefited greatly from access to microfinance and were consequently increasing their crop production. It was interesting to see in this project that the provision of microfinance was part of an integrated development package. I am aware that microfinance is often provided in agricultural development projects as a stand-alone input, and that this does not always provide the expected benefits. The prospect of these loans having meaningful benefits seems much more likely when, as in this project, they are given together with assistance to develop better farming systems and new facilities, as well as training and other institutional support.

The final component of the project was to provide assistance and training in the establishment of village committees to plan and implement small community infrastructure development projects. The community has to meet part of the cost of these projects, so creating a situation where they take responsibility for raising and using funds wisely. For example, several communities had installed water standpipes in their communities, bringing potable water in pipes from hillside springs 2–10 km away. This clean water had greatly improved the health of community members, as well as eliminating the need for women to carry water from contaminated streams and rivers. The water was also being used for livestock, village nurseries and small-scale irrigation of off-season vegetable plots. Two communities had also used profits from their nursery to help pay for the digging of wells.

At Batibo, one of the villages to recently join the project, the community was in the process of building a bridge across the local river and creating a road to an area of forest that they wished to open up to agriculture. I was invited to attend a community council meeting in the Fon's palace, presided over by His Majesty the Fon. It was a very formal meeting. The project was explained to me in great detail by one of the council members who is a lawyer. I was then asked to comment on what I had seen. This gave me the opportunity to also explain how agriculture could be implemented in ways that would provide food and yet maintain forest habitat for wildlife. This was of great interest to the council, as one of their concerns was that while the road had many local benefits it might lead to loss of forest and particularly to the loss of animals important for bush meat. The importance of bush meat became very clear when after the meeting I was treated to lunch and tree pangolin was on the menu! I have faced this dilemma before: on the one hand you feel that you have to eat the food put in front of you out of respect for your hosts, and on the other you don't want to eat something that you would rather see running around in a forest.

The most important and exciting thing about this development project was the wide range of positive livelihood impacts that the farmers reported to us (Box 11.1) and that seemed to be truly transforming people's lives (Fig. 11.3). Taken together, these impacts illustrate that better land husbandry and crop diversification create productive farmland, improve the rural economy and deliver a number of social benefits, including increased engagement in a self-help philosophy to community welfare and advancement. If we let our imagination take us still further, the large-scale development of new businesses and employment could lead to a restructuring of society in which a high proportion of the population would be engaged in economic activity. This diversification of livelihood options would get society away from the current situation where about 80% are smallholder farmers. This diversification of the rural economy seems to me to be an essential component of economic development in the tropics.

These impacts strongly suggest that by promoting self-sufficiency through the empowerment of individuals and community groups through the provision of new skills in agroforestry, food production and processing, community development and microfinance, it is possible for communities to climb the entrepreneurial ladder out of poverty. By so doing, they set themselves on a path towards improved livelihoods based on better nutrition and the recognition of the social and cultural value of 'life-support systems' of indigenous species formerly ignored by agricultural science. Through this project, rural people are gaining a vision of a new future with better livelihoods and fewer hardships – or at least different hardships. While the target group here is the rural poor, there are indirect benefits to the urban poor too, as indigenous fruits, nuts, medicines, etc. are sold in urban markets They are much cheaper than supermarket products.

To illustrate this, one of the very encouraging outcomes of these nursery developments, attributed to the income-generating capacity of these

Box 11.1. Impacts reported by farmers involved in the Food for Progress programme in Cameroon.

Positive impacts

- Increased number of farmers adopting agroforestry and the domestication of indigenous trees.
- Increased production of tree products.
- Increased income from tree sales by nurseries.
- Increased income from sale of tree products.
- Increased income from better farming practices.
- Increased income from eligibility for microfinance.
- Increased income used for schooling and school uniforms.
- Increased income used for medicines and health care.
- Increased income used for home improvements (e.g. installation of water and electricity in the home, new buildings, etc.).
- Increased income used for farm improvement (e.g. livestock, wells, agricultural inputs, better nurseries, etc.).
- New employment opportunities from nurseries. Currently there are 123 agroforestry nurseries.
- New employment opportunities from processing both agricultural crops (such as cassava) and new markets for processed agroforestry products (fruits, spices, herbs and medicinal products).
- New employment opportunities in the emerging workshops producing small tools and appropriate mechanized equipment to service the need for food processing equipment.
- New employment opportunities from marketing as traders of processed products and the food processing equipment.
- New employment opportunities in transport from producers to markets and to the processors of agricultural produce.
- Retention of youths in the villages due to career opportunities by domesticating trees in their village nurseries. A few youths have returned from the town to their villages. Some are setting up commercial nurseries.
- Tree domestication has led to better diets and improved nutrition. People used to eat corn fufu and some vegetables but now they are eating fruits and more vegetables. Livestock fed on tree fodder has also led to the consumption of more meat.
- Luxury food items consumed. One lady described having honey and herb tea for breakfast as a luxury.
- Improved health from potable water. In addition there were benefits for livestock health and production.
- Piped water supplies for irrigation and use in nurseries.
- Increased livestock rearing due to tree fodder.
- Increased use of traditional medicines and better health.
- Increased honey production and processing.
- Reduced drudgery in women's lives from not having to collect water from rivers and farm produce from remote farms, as well as from mechanical processing of food crops.

Continued

> **Box 11.1.** Continued.
>
> - Reduced drudgery gives more time to look after their families and engage in farming or other income-generating activities.
> - Improved marketing for food and agroforestry products.
> - Improved soil fertility from improved fallows has led to a threefold increase in maize yield with additional benefits from improved weed control.
> - Improved tree fodder for goats and cattle.
>
> As a result of better farming methods farmers had more time for marketing and new farming activities. The community feels empowered, stronger and more optimistic for the future as they have learned new ways of farming that they could sustain. Villagers from other villages were turning to them for help and advice. One group of people said that if they continued down this path they felt sure they could aspire to catch up with people that previously they had felt to be of unattainably higher status than they were themselves. Knowledge has empowered the Rural Resource Centres (RRCs) as an agent of change. Many people (100% of those interviewed) acknowledged that through the training provided by the project that they had gained knowledge and understanding of agroforestry, tree domestication, business practices and management, and in team work and that this knowledge was changing their lives.
>
> **Negative impacts**
>
> - Increased theft.
> - Increased jealousy.
> - New roads lead to deforestation and land degradation as a result of the expansion of farming activities to more remote areas.

nurseries, has been the retention of youths in the villages. At PROAGRO, for example, about 10% of the 295 participating farmers in 16 nurseries were youths who had decided that they could now have better lives in the villages than in the towns. We even heard of one young man who had returned from the town to live in the village. To recognize the importance of this outcome, perhaps we need to remind ourselves that young people often go to the towns in search of employment to support their families back in the villages. However, the dream of employment is frequently unfulfilled and these youths live in slums and turn to crime and drugs.

I must not give the impression that every farmer in the project area was an adopter. Rates of adoption in agriculture are often low, as farmers are careful not to start something until they are sure it will meet their needs and situation. Understandably, when the ability to feed one's family from a very small area of land is at stake, farmers do not want to take unnecessary risks. When Zac Tchoundjeu's team looked at the rate of adoption, they found a remarkably good uptake of the tree domestication activities, ranging from 55 to 89% in different areas. Again not surprisingly, those who dropped out were found to be people with secondary education and in other forms

Fig. 11.3. Farmers of the Rural Resource Centre (RRC) of RIBA in Bui in Cameroon expressing their enthusiasm for agroforestry and tree domestication, which have transformed their lives.

of employment (mainly civil servants), who basically were only in their villages at the weekend. The highest rate of adoption was among middle-aged men and they indicated that their motivation was income generation and income diversification. In no small part, I think, the high adoption rate is because the process builds on local knowledge, tradition and culture, builds on local markets, and is clearly targeted at delivering the benefits that the farmers themselves identified at the start of the process.

When I got back home from Cameroon, I was excited about the progress that had been made in this project. The dreams of these farmers had come true – their incomes had risen from around US$1/day to US$5–15/day. This doesn't sound much to us in industrialized countries, but in rural Cameroon this is significant – indeed well above the targets set by the UN Millennium Development Goals. The amazing thing about this is that it was not technically difficult to achieve this enormous success.

Pulling this together, we have to remember that several international reports have said that 'business as usual' is not an option for the future. What we have seen in this chapter is that the early signs of scaling up tree domestication indicate that the agroforestry model for multifunctional agriculture does seem to work in practice. Is the birth of a new approach to agriculture and rural development in the tropics a new paradigm for poverty alleviation, reduced malnutrition and hunger, and the sustainable use of natural resources? Although the outcomes of the Food for Progress project are miniscule in relation to the scale of the global problem, the implications for future progress are huge. Just as a germinating seed can become part of a future forest, I believe this project is a significant leap towards, and a model for building, a better future for smallholder farmers in Africa and elsewhere.

If applied on a large enough scale, such agroforestry initiatives within an integrated rural development programme would also significantly increase carbon sequestration in woody biomass and in soil carbon, and protect watersheds, as well as enhancing biodiversity and better agroecosystem function. In other words, we would be addressing many 'pressure points' in the cycle of land degradation and social deprivation in the tropics (Fig. 2.1).

Obviously, it is true that the success stories that I have mentioned have been achieved on only a very small scale so far; like a drop in the ocean. If we want to change the world for the better, the challenge is how to scale up all these small steps forward to have meaningful impact. This challenge is a logistical one – how to scale up from a few thousand farmers to millions, or better still billions of poor farmers around the world? Work continues to bring all this together, but the scaling up is dependent on a greatly increased determination by the global community to make multifunctional agriculture a reality, especially in the tropics.

The further scaling up of participatory domestication to tens of millions of new households every year across the developing world is a huge challenge, in terms of both the logistics of training and supervision and the adaptation to new species, environments and markets. In this connection, there is an urgent need to rapidly expand the pool of expertise in techniques like vegetative propagation. Fortunately, the Commonwealth Science Council published an excellent manual by Alan Longman in 1993,[2] with supporting videos by the Edinburgh Centre for Tropical Forests. A tree domestication training course based on the contents of this chapter is now available as narrated PowerPoint slides on the ICRAF website.[3]

I hope I have sown a few seeds that can grow and multiply. To illustrate the potential for meaningful impact, recent studies have found that there are already over 1 billion ha of agricultural land with 10% tree cover, 80% of which is outside Europe and North America. This area is about 46% of the world's cultivated land, so the expansion of agroforestry technologies across this area should have dramatic benefits on the lives of the millions of people already living there. Some development agencies, governments and large investors consider this to be useless 'marginal land' of no particular value and so promote its conversion to plantation monocultures. Hopefully this book explains why this is a serious mistake, as it is land that can be easily rehabilitated and converted to productive agroforests of great value to local people.

Finally, Food for Progress is a community development project that has hit the right button, or perhaps more accurately the right set of buttons. In 2010, I heard that one of the RRCs in this project had won the prestigious Equator Prize sponsored by the United Nations Development Programme, the private sector, civil society and governments, to raise the profile of grassroots efforts to reduce poverty. Coming on top of being selected as one of Africa's success stories by the Foresight Global Food and Farming Futures Project of the UK's Government Office for Science, this project is starting to be recognized as a really important and innovative approach to rural development.

One important aspect of this project is that its success has been achieved by the provision of knowledge, support and mentoring, and not by handouts of money to the communities. In my experience of development projects, throwing money at communities is not the solution, as the projects collapse when the money dries up. This project provided the villagers with expertise so that they could empower themselves. I very much hope that the participatory, self-help approach of the Food for Progress project will give it the capacity to be self-sustaining when the development funds dry up.

Notes

[1] Provided by CANADEL (Centre d'Accompagnement de Novelles Alternatives de Développement Local), FIFFA (First Investment for Financial Assistance) and Winrock International, respectively. Ebenezar Asaah was the overall project manager.
[2] Longman, K.A. (1993) *Rooting Cuttings of Tropical Trees.* Propagation and Planting Manuals, Volume 1. Commonwealth Science Council, London.
[3] www.worldagroforestry.org/Units/training/downloads/tree_domestication

Further Reading

Asaah, E.K., Tchoundjeu, Z., Leakey, R.R.B., Takousting, B., Njong, J., *et al.* (2011) Trees, agroforestry and multifunctional agriculture in Cameroon. *International Journal of Agricultural Sustainability* 9, 110–119.
Leakey, R.R.B. (in press) Addressing the causes of land degradation, food/nutritional insecurity and poverty: a new approach to agricultural intensification in the tropics and sub-tropics. In: Hoffman, U. (ed.) *UNCTAD Trade and Environment Review 2011/2012.* United Nations Conference on Trade and Development (UNCTAD), Geneva. Available at: www.unctad.org/Templates/Page.asp?intItemID=3723&lang=1 (accessed 6 February 2012).
Leakey, R.R.B. and Asaah, E.K. (in press) Underutilized species as the backbone of multifunctional agriculture – the next wave of crop domestication. *Acta Horticulturae.*
Tchoundjeu, Z., Degrande, A., Leakey, R.R.B., Simons, A.J., Nimino, G., *et al.* (2010) Impact of participatory tree domestication on farmer livelihoods in West and Central Africa. *Forests, Trees and Livelihoods* 19, 219–234.

The Convenient Truths   12

> How could a society which was once so mighty end up collapsing? ... Might such a fate eventually befall our own wealthy society? Will tourists someday stare mystified at the rusting hulks of New York's skyscrapers, much as we today stare at the jungle-overgrown ruins of Maya cities?
>
> Jared Diamond (2005) *Collapse: How Societies Choose to Fail or Survive*. Penguin Group, Camberwell, Australia.

> It is difficult to get a man to understand something when his salary depends upon his not understanding it.
>
> Upton Sinclair quoted by Al Gore (2006) in *An Inconvenient Truth*.

In his book *An Inconvenient Truth*, former US Vice President Al Gore[1] spelled out the need to address climate change, but climate change is only one of the problems arising from agriculture and our mismanagement of global resources. By contrast, the 'convenient truth', as I have tried to illustrate in this book, is that we do have some knowledge and experience of how we can improve our approach to agriculture in ways that can help to mitigate the consequences of our profligate lifestyle.

It is not easy to unravel the impacts of modern agriculture from manufacturing and urban consumerism, because modern agriculture is highly dependent on energy consumption for mining ores, transportation and the production of machinery, farm buildings and agrichemicals such as fertilizers and pesticides. Nor is it particularly useful to try to unravel these things, as the Industrial Revolution is here to stay. Instead we just have to make sensible adjustments to our ways of doing business; we need to be more aware of the needs of others and we need to improve the way we practise agriculture. We are not talking about dramatic changes that will leave people worse off; as much as anything it is a change in attitude towards a more integrated approach to world development. Just as the problems we have created are interconnected, so are the solutions, and lots of small positive changes being multiplied by numerous interactions can result in major impacts. This reminds me of an African proverb:

> If many little people,
> in many little places,
> do many little things,
> they will change the face of the world.

I think there is a lot of truth behind this proverb, but if lots of little people are going to do lots of little things, they all need to be pulling in the same direction. Unfortunately, despite the wonders of the Internet, our world is full of 'disconnects' between groups of people who ought to be in communication. Possibly it is the surfeit of knowledge and information that leads to this. It means we cannot cope with the quantity of information in our own areas of involvement, let alone that from other disciplines and professions.

The existence of these disconnects is particularly a problem when we are trying to address as complex a problem as the interlinkages between: (i) use of natural resources; (ii) land degradation; (iii) extraction and pollution of groundwater; (iv) loss of biodiversity and ecosystem function; (v) climate change; (vi) poverty; and (vii) hunger and malnutrition, especially in the tropics and subtropics (Fig. 2.1). We see the failure to connect these issues nearly every day when we read the newspapers or listen to the news. Politicians, policy makers and the media continuously discuss each individual issue as if it functioned in isolation from all the rest. In the last chapter I suggested that multifunctional agriculture as part of a programme of integrated rural development is one way of bringing together many of the critical drivers of change and making progress on many different fronts. It seems to me that this approach also makes much better use of our limited funds. So, I believe we should be seeking to address issues of poverty, malnutrition, land degradation and climate change simultaneously, and not in isolation from one another. Isolationism will lead to misconnections and the duplication of resource use, not to mention undue bureaucracy.

So, to find a way forward towards more sustainable tropical agriculture and rural development, let's try to pull some key messages together. I believe we can identify five convenient truths from all that we have learnt in the earlier chapters.

Convenient Truth 1

We have nearly 30 years of research and development experience in the use of well-tested, appropriate and adoptable soil improvement technologies ready for wider application. These technologies are appropriate for the rehabilitation of degraded soils and the improvement of food security. They would also greatly increase the ability of farmers to support their families on small areas of land. If more widely applied, they would increase the returns from the investment in the Green Revolution. Rehabilitation on a scale to provide sufficient productive land to meet the needs of smallholder farmers would have the added benefit that it should eliminate the need for further deforestation for agriculture. This would be a positive contribution to increased food security and to minimizing climate change.

Convenient Truth 2

We have nearly 15 years of research and development experience of how to domesticate traditionally important indigenous trees as new cash crops, yielding a wide range of marketable and traditionally important products with which we can diversify, enrich and intensify farming systems and rebuild the cultural psyche of indigenous peoples. We can empower poor farmers to be self-supporting by the practice of participatory tree domestication based on appropriate technology, increased knowledge and adoptable skills that allows further steps to food security, environmental rehabilitation and reduction of climate change. It also opens up a pathway to higher living standards for themselves and their children. The enhancement of livelihoods, income generation, rural employment and economic growth come about because:

- The research that has been done has greatly improved understanding of the scientific principles of vegetative propagation that will allow us to rapidly and simply domesticate tropical tree species virtually unknown to science. Consequently, we now have robust, simple techniques of vegetative propagation for tropical trees which are valued and now being used by farmers and local communities to develop their own cultivars.
- Research has been done to characterize the tree-to-tree genetic variation in a wide range of traits conferring market value to many different agroforestry tree products. This has provided the framework for genetic selection and the domestication of new tree crops.
- We have experience of creating new agroforestry tree crops, which will start to generate income in 3–4 years. At this stage, the need is not to produce tens of thousands of US dollars. Just US$1000 would be a quantum leap for hundreds of millions of poor farmers and could be the springboard to a new and better life. These new crops are compatible with and adapted to the local environment and local people's needs. Some of the products from these new crops will be improvements on wild harvested products for home consumption and domestic use, while others have local, regional and international markets.

They will have better quality and greater uniformity, attributes that are recognized by customers and result in higher market prices and greater demand. The products from other cultivars will be produced 'out of season', so expanding the season of production and increasing market value. Some new international commodities are already being developed using these tree domestication techniques.
- We have some experience in commercializing the products of these new tree crops, but much remains to be done. Nevertheless at the local level domestication has proved to be a stimulus to entrepreneurism in value adding and trade. Recent evidence indicates that this leads to the creation of opportunities for employment and further income generation. It is envisaged that this can in turn lead to greater financial independence from agricultural production, so paving the way for a society that has reduced dependency on everyone being a farmer. Women in particular are involved in local trade. Interestingly, young people are seeing new opportunities to engage in commerce and to remain in their villages rather than migrating to local towns and cities.
- Local people have great interest in the species producing traditionally important foods and medicines and other day-to-day products, many of which are also important in cultural festivals recognizing ancient customs, folklore and beliefs, as well as protecting the spirits of tribal ancestors. Consequently, this approach brings together agricultural science and technology with traditional knowledge. Importantly, many of the products from these new crops are rich in micronutrients, protein and oils and so their increased consumption by households with malnutrition and poor nutritional security should lead to a more balanced diet, with its associated health benefits arising from an enhanced immune system.

Convenient Truth 3

We know how to use the three steps of agroforestry to initiate and implement integrated rural development projects that reduce the yield gap between potential and actual yield in staple food crops, through the combination of land rehabilitation, the domestication of new crops and the commercialization of their products.[2] Evidence is emerging of the start of a flow of positive impacts at the community level when the three steps of agroforestry form the skeleton of an integrated rural development programme like the Food for Progress project. The issue to be resolved is how to scale up to the point that the social, economic and environmental benefits are visible on national, regional and global scales. This comes about because:
- Households practising participatory tree domestication have been found to be using their newly generated income to: (i) improve their

farms (to develop commercial tree nurseries, purchase livestock, build wells and improve buildings, install electricity); (ii) improve village infrastructure (pumps for potable water supplies, bridges, store houses); (iii) send children to school; and (iv) pay for health care. The greater availability of water in communities, and better access to fields and markets, means that the drudgery of women's lives is being reduced, and consequently they have more time to look after their families and to work on their farms.
- Evidence indicates that the diversification of farms with new crops (planned biodiversity) leads to better land husbandry and to species-rich agroforests (unplanned biodiversity) that become fully functional mature agroecosystems, which are also attractive habitats for wildlife, as well as reducing soil erosion and protecting watersheds within a productive and attractive landscape. The income from an increased number of species creates a cash flow at different times of the year increasing financial resilience to crop failure. These systems therefore are appropriate for farmers with no off-farm income from investments, pensions, social security or insurance.
- Very importantly for poor farmers without opportunities for social services in hard times and who have to be self-sufficient and support their families on a small area of land, this diversified production system is risk averse and empowering. It confers social and economic sustainability and allows economic growth.
- Agroforests containing a wide range of woody perennial species reduce the emissions of carbon dioxide and other GHGs and so help to mitigate climate change.[3]

Convenient Truth 4

The recent involvement of some multinational companies in integrated rural development projects using agroforestry technologies and tree domestication is contributing to the up-scaling of these approaches to multifunctional agriculture. Such changes represent a new paradigm for commerce in developing countries and conform to the concept of 'enlightened globalization'. It is foreseen that this new paradigm of PPP should in the longer term have economic benefits for both the developed country partners and the industries.

Convenient Truth 5

Recent evidence shows the scale of tree-based farming systems to be over 1 billion ha in many of the most impoverished parts of the world, which are also occupied by some of the poorest people. This indicates the enormous opportunity to add agroforestry technologies that would improve the productivity and ecological resilience of farming systems. These 1 billion ha are a potential starting point for a large-scale initiative to implement agroforestry as a delivery mechanism for multifunctional agriculture. Past experience would suggest that farmers living in these areas would be enthusiastic participants and thus that there is a good chance of adoption on a scale that would have real impact at the level of the global problem.

This set of five convenient truths illustrates that the challenge is not technically difficult, as the overall approach is basically simple, readily adoptable, relatively inexpensive and not dependent on high-technology. The problem is one of logistics and implementation on a scale to have meaningful and positive socio-economic and environmental impacts on a global level.

It really is just a matter of finding the political will to do the job and give the poor and disadvantaged their rights to access better livelihoods through agricultural development. Thinking about the topsy-turvy social order of modern life, we pass laws against discrimination, we abhor prejudicial behaviour and say we support equality, but then we implement repressive economic policies and support commercial policies and trade that subjugate primary producers overseas. So really we live a fallacy. Our actions as consumers speak more loudly than words when we in the industrial countries demand a wide range of food and non-food products and implement repressive trade policies over the producers.

The test for humanity is to find the political will to rise to this challenge, which runs counter to our modern egocentric culture. Despite our politically correct statements, we have a bad track record of finding ways to create a fairer world – a world where all people, regardless of race, creed, ethnicity, nationality or geographical region, are equally important.

I have restricted these convenient truths to tropical agriculture, but there is also growing interest in agroforestry in temperate and Mediterranean countries, but these areas are different in many ways, and I don't want to confuse my message by addressing them. The IAASTD reports give a much broader and more comprehensive set of information about how to redirect agriculture towards more sustainable rural development worldwide.

So, you may be asking, why has the agroforestry approach to multifunctional agriculture not happened before now? Well, actually, much of it has been around for at least 100 years, as we saw in Chapter 3. However, for wider scale application there have been at least two missing ingredients in these precursors to multifunctional agriculture. The first missing ingredient was technical knowledge about tree domestication, although as we saw in

Chapter 6 the farmers did initiate this too, but they lacked the knowledge of the necessary techniques of vegetative propagation to go the fast route and develop cultivars. Consequently they were left dependent on the very slow route of planting selected seedlings and waiting at least 10 years to find out if any of these were better than average.

The second missing ingredient was appropriate policies. This sad state of affairs arose because the colonial masters did not recognize the importance of local trees, except perhaps for timber, and even these were seldom planted. Instead the policy focus was on clearing land and planting staple food crops or a small number of cash crops for the export market. This has been the conventional wisdom ever since, and still persists in many quarters.

This book is the culmination of 30 years of work on tree domestication and agroforestry. I hope that it will be the spark to ignite a revolution based on the above convenient truths – a revolution that we first dreamed about in 1992 when we organized the conference 'Tropical Trees: the Potential for Domestication and the Rebuilding of Forest Resources'. Interestingly, the process of domesticating species to make them more useful to mankind has recently been accredited as being a powerful stimulus in the rise of civilizations. As we have seen earlier, Jared Diamond[4] in *Guns, Germs and Steel* has said that food crop domestication has been 'the precursor of settled, politically centralized, socially stratified, economically complex and technologically innovative societies'.

Exactly what the trigger is for domestication is not clear, but some attribute it to the rise of demand exceeding the prevailing supply. Clearly from much of what we have seen earlier, this has mostly impacted on what we call the developed countries of mainly temperate and Mediterranean zones. With regard to the current generation of food crops, the domestication of several cereals (barley and wheat) was initiated in the Neolithic Age, around 8000 BC in the Near East, while the process in rice was in the Far East at about the same time. Interestingly, the domestication of the first tree crops (dates, olives and grapes) can be traced back to about the same time, again in the Near East. The domestication of other tree crops, like figs, oranges and apples, goes back to about 3000 BC in the Near East, China and Central Asia, respectively. The domestication of sheep, pigs and cattle started at about the same time as wheat and barley, but was preceded by goats (10,000 BC) and dogs (15,000 BC).

Historically, the main centre of domestication in the tropics was the Highlands of Papua New Guinea, where according to Jared Diamond there is archaeological evidence of farming systems about 6000 years ago, which probably involved the domestication of species such as taro, yams and bananas.

One of the surprising aspects of crop domestication until very recently is the small number of species involved – only 0.5% of all edible species (0.04% of all higher plants). In recent decades there have been relatively few new domesticates. Among the woody species the two most obvious are probably the kiwi fruit (*Actinidia chinensis*) and the macadamia nut (*Macadamia ternifolia*). To a considerable extent, this can be attributed to the high research, development and marketing costs, for which I have not been able to find even a ballpark figure in the literature. At first sight therefore it seems that by advocating the domestication of agroforestry trees as a critical component of developing multifunctional farming systems, I am suggesting a very expensive approach to delivering real progress towards the alleviation of poverty, malnutrition, hunger and environmental degradation. However this is not the case, as I believe the participatory approach to domestication in rural communities is much less expensive than the Green Revolution model of crop breeding, as it is less sophisticated and there is much less laboratory work.

Inevitably, this raises the question of the relative costs and the benefits, and here it is very difficult to be precise. This is not the place for an analysis of the relative costs, but certainly I think one should be done so that an informed decision could be made about the initiation of a new wave of domestication, this time focused on tropical species which provide the everyday needs of smallholder farmers of developing countries. This could lead to a new round of economic development in countries where currently business, employment and trade are dominated and constrained by foreign values, standards and interests.

We have seen that in the past, despite best intentions, errors have been made in guiding the evolution of agriculture. Obviously, this risk still exists, and no doubt there are flaws in the ideas that I have presented based on current experience of agroforestry. One such risk is that the domestication of indigenous fruit and nut trees could be so successful that an entrepreneur or company decides to develop monocultural plantations of some of the new indigenous trees domesticated by rural communities, perhaps in an overseas location with a similar climate. This would undermine the whole purpose of developing the new crops as a means to enhancing food security, health, income generation and environmental rehabilitation in Africa and other developing countries.

At the moment the risk of this does not seem too great, as the market demand is mostly at the level of local, rather than global, markets. At this level, I believe the potential benefits from domestication will outweigh the risks. Nevertheless, to further reduce these risks there is a need for the international community to rapidly resolve the issues around the development of a new form of intellectual property that protects the innovations of local communities developing cultivars from their indigenous trees. The need is for something akin to plant breeders' rights, which will protect the innovations of poor farmers from exploitation by unscrupulous entrepreneurs.

Throughout this book, I have been saying that we must strive to move agriculture forward in ways that advance the lives and economies of all the peoples and countries of the world, without losing their unique cultures and traditions, unique environments and vegetation and their associated biological diversity. I have tried to present some pointers to how we can develop more sustainable approaches to agriculture, especially in the tropics where the needs are greatest. I have been suggesting a more integrated form of agriculture, one that builds on the advances of the Green Revolution and one that, through agroforestry, adds greater attention to land husbandry and the overlooked Trees of Life. What we see emerging here is an integrated approach to resolving many of the world's big social and environmental problems. This seems to make sense when dollars are in short supply.

Much of what I have presented is poorly known outside the agroforestry community, and so may be new to you. Much of what I have presented also runs counter to conventional wisdom; wisdom that many academics and other agricultural professionals are challenging. I have presented my personal view based on my experience. I believe it is possible to meet Julian Cribb's exacting and impossible-sounding requirements for the future of agriculture – to double food output – 'using far less water, less land, less energy and less fertilizer'. I hope this book highlights a way forward towards a better future for billions of poor people, based on the Trees of Life. To move towards a new 'Eden' will require public interest and political will. You can help to make a difference by raising the profile of the issues we have examined in this book. The good news is that people are starting to raise many of the issues in public forums.

I recently went to Nairobi to sit on a panel at a special event entitled 'Agriculture and Food Security: Will the next revolution be more sustainable?' during the 25th meeting of the United Nations Environment Programme (UNEP) Governing Council. The agenda of this special event was to discuss the findings of IAASTD vis-à-vis the 'environmental food crisis'. At this meeting, Christian Nelleman of UNEP presented a talk on 'the Environment's Role in Averting Future Food Crises'. One of the points in his seven-point action plan was:

> Support farmers in developing diversified and resilient agriculture systems that provide critical ecosystem services (water supply and regulation, habitat for wild plants and animals, genetic diversity, pollination, pest control, climate regulation) as well as adequate food to meet local and consumer needs.

This is what I have been talking about in this book. We have also covered several of the other issues he raised, such as developing 'climate-friendly agricultural production systems and land-use policies at a scale to help mitigate climate change'.

In the discussion session after the presentations at this meeting, Professor Wangari Maathai, winner of the Nobel Peace Prize in 2004, asked if Africa would be able to feed itself in the future. My response was: 'Certainly, yes, it can, if we implement multifunctional agriculture on a meaningful scale.' Coming from her this question is especially important, as she has clearly highlighted the problems of her continent in her book *The Challenge for Africa*.[5]

Notes

[1] Al Gore (2006) *An Inconvenient Truth: the Planetary Emergency of Global Warming and What We Can Do About It.* Rodale, Emmaus, Pennsylvania, 325 pp.
[2] As we saw in Chapter 10, genetically modified staple food crops could also contribute to achieving Step 1.
[3] Recent studies by the World Agroforestry Centre suggest that carbon sequestration could be increased from 2.2 up to 90–150 t of carbon/ha by the establishment of agroforests in areas currently degraded by agriculture. Currently some 900 million ha worldwide could be ameliorated in this way.
[4] Jared Diamond (1999) *Guns, Germs, and Steel: the Fates of Human Societies.* WW Norton & Co., New York, 494 pp.
[5] Wangari Maathai (2009) *The Challenge for Africa.* Arrow Books, London, 319 pp.

Postscript

Since preparing the manuscript of this book, I have become Vice Chairman of the International Tree Foundation (ITF), a UK registered charity. ITF has promoted and funded sustainable community forestry projects in the UK and overseas for almost 90 years and has been responsible for the establishment of hundreds of millions of trees.

The primary roles of ITF are:

- the funding of tree-planting projects in the UK and overseas, particularly in Africa;
- advocacy and the promotion of public awareness of global issues around deforestation, reforestation and agroforestry; and
- the implementation of development projects, in partnership with local communities and international agencies, that apply research outputs to enhance the livelihoods of the rural poor.

ITF projects are designed around local needs and aspirations, and carried out in partnership with local community-based organizations in ways very closely aligned to the content of this book. We would greatly appreciate your support. Please visit our website (www.internationaltreefoundation.org).

Appendix: Author's Experience Prior to the Events of this Book

From 1974 to 1993, the author was Senior/Principal Scientific Officer at the Natural Environment Research Council's Institute of Terrestrial Ecology (formerly the Institute of Tree Biology) near Edinburgh, Scotland. From 1987 he was Project Group Leader (Tropical Forestry and Mycorrhizas Group) responsible for ten research scientists, two technicians and 11 postgraduate students. The mission of this research team was to promote sustainable management of renewable natural resources and the environment in the tropics. Projects were established in Costa Rica, Trinidad, Brazil, Chile, Senegal, Ghana, Cameroon (×2), Kenya (×2), Malaysia and Indonesia with the objectives of:

1. Promoting the restoration and conservation of tropical forests by studies on the ecology and regeneration processes of forest and woodland ecosystems in the moist and dry tropics.
2. Undertaking multidisciplinary strategic research involved in the development of techniques for:
   (a) rapid domestication of tropical trees for a range of forest products, to provide renewable and more productive resources for future use;
   (b) establishment of trees for sustainable forestry and agroforestry, in ways that enhance biodiversity and ecosystem stability; and
   (c) rehabilitation of both degraded forests/woodlands and agricultural land, by the successful establishment of trees with their associated root microsymbionts, so restoring damaged ecosystems.
3. Undertaking strategic research on the physiology and growth of trees, especially:
   (a) the capture, analysis and exploitation of genetic variation; and
   (b) the interactions between genetic variation and the environment.
4. Undertaking strategic research on the physiology and epidemiology of components of the symbiotic root microflora, especially the role of mycorrhizas in promoting tree performance.

5. Developing techniques to increase the productivity of man-made ecosystems, while sustaining ecological integrity.
6. Attaining an understanding about the functioning of ecosystems and the 'knock-on' effects that lead to ecosystem failure.

In addition, the author undertook consultancy work for:

1. United Nations Food and Agriculture Organization (1982) in support of the Seed Sources and Tree Improvement Project, Forest Research Centre, Sandakan, Sabah, Malaysia.
2. UNEP and UNESCO (1983) to organize a regional workshop to discuss plans for a West African Regional Hardwood Improvement Programme.
3. European Development Fund DG VIII (1984) to develop a Regional Programme for the Improvement of Tropical Hardwoods in West and Central Africa (Sierra Leone, Liberia, Ivory Coast, Ghana, Nigeria, Cameroon and Congo).
4. World Bank Forestry Project in Republic of Cameroon (1985) with Office National de Regénération des Forêts (ONAREF) to make recommendations for a Tree Improvement and Seed Production Unit in the four ecological zones of the country from rainforests to the Sahel.
5. UNDP/FAO Interagency Forestry Sector Review of Cameroon (1987) as forest genetics and seed supply specialist.
6. World Bank Forestry Project in Cameroon (1987, 1989, 1990, 1991, 1992) to initiate ONAREF's Tree Improvement and Seed Production Unit.
7. UK Overseas Development Administration (1988, 1989, 1990, 1991, 1993) to develop and implement proposals for a project at Centro Agronómico Tropical de Investigación y Enseñanza (CATIE), Costa Rica on Vegetative Propagation and Tree Improvement for Tropical Agroforestry.
8. UK Overseas Development Administration (1989) to develop proposals for a project on vegetative propagation of Mysore Gum at Mysore Paper Mills, Shimoga, India.
9. Ghana Ministry of Lands and World Bank Forest Resource Management Project at Forest Products Research Institute, Kumasi, Ghana.
10. UK Overseas Development Administration (1992) to prepare research plans for Tropical Forest Management Project in Central Kalimantan, Indonesia.
11. UK Overseas Development Administration and UK Natural Resources Institute (1993) to provide advice and training on the propagation of indigenous trees in Malaysia.
12. Smith & Salleh (1993) to provide advice on the rehabilitation of hill dipterocarp forest in Johore State, Malaysia.
13. Australian Meat Research Commission (1996) to review and evaluate research on tagasaste (tree lucerne) and to define research priorities.

Further Reading

Ladipo, D.O., Grace, J., Sandford, A. and Leakey, R.R.B. (1984) Clonal variation in photosynthesis, respiration and diffusion resistances in the tropical hardwood tree *Triplochiton scleroxylon* K. Schum. *Photosynthetica* 18, 20–27.

Ladipo, D.O., Leakey, R.R.B. and Grace, J. (1991a) Clonal variation in apical dominance in young plants of *Triplochiton scleroxylon* K. Schum.: responses to decapitation. *Silvae Genetica* 40, 135–140.

Ladipo, D.O., Leakey, R.R.B. and Grace, J. (1991b) Clonal variation in a four year old plantation of *Triplochiton scleroxylon* K. Schum. and its relation to the predictive test for branching habit. *Silvae Genetica* 40, 130–135.

Ladipo, D.O., Leakey, R.R.B. and Grace, J. (1992) Variations in bud activity from decapitated, nursery-grown plants of *Triplochiton scleroxylon* in Nigeria: effects of light, temperature and humidity. *Forest Ecology and Management* 50, 287–298.

Ladipo, D.O., Britwum, S.P.K., Tchoundjeu, Z., Oni, O. and Leakey, R.R.B. (1994) Genetic improvement of West African tree species: past and present. In: Leakey, R.R.B. and Newton, A.C. (eds) *Tropical Trees: Potential for Domestication. Rebuilding Forest Resources.* HMSO, London, pp. 239–248.

Leakey, R.R.B. (1986) Prediction of branching habit of clonal *Triplochiton scleroxylon*. In: Tigerstedt, P.A., Puttonen, P. and Koski, V. (eds) *Crop Physiology of Forest Trees.* University of Helsinki, Finland, pp. 71–80.

Leakey, R.R.B. (1987) Clonal forestry in the tropics – a review of developments, strategies and opportunities. *Commonwealth Forestry Review* 66, 61–75.

Leakey, R.R.B. (1991) Clonal forestry: towards a strategy. Some guidelines based on experience with tropical trees. In: Jackson, J.E. (ed.) *Tree Breeding and Improvement.* Royal Forestry Society of England, Wales and Northern Ireland, Tring, UK, pp. 27–42.

Leakey, R.R.B. and Ladipo, D.O. (1987) Selection for improvement in vegetatively-propagated tropical hardwoods. In: Atkin, R. and Abbott, J. (eds) *Improvement of Vegetatively Propagated Plants.* Academic Press, London, pp. 324–336.

Leakey, R.R.B. and Longman, K.A. (1986) Physiological, environmental and genetic variation in apical dominance as determined by decapitation in *Triplochiton scleroxylon*. *Tree Physiology* 1, 193–207.

Leakey, R.R.B., Ferguson, N.R. and Longman, K.A. (1981) Precocious flowering and reproductive biology of *Triplochiton scleroxylon* K. Schum. *Commonwealth Forestry Review* 60, 117–126.

Longman, K.A., Leakey, R.R.B. and Denne, M.P. (1979) Genetic and environmental effects on shoot growth and xylem formation in a tropical tree. *Annals of Botany* 44, 377–380.

Longman, K.A., Manurung, R. and Leakey, R.R.B. (1990) Use of small, clonal plants for experiments on factors affecting flowering in tropical trees. In: Bawa, K.A. and Hadley, M. (eds) *Reproductive Ecology of Tropical Forest Plants.* Man and the Biosphere Series, UNESCO Paris and Parthenon Publishing, Carnforth, UK, pp. 389–399.

Newton, A.C., Baker, P., Howard, W., Ramnarine, S., Mesén, F.J. and Leakey, R.R.B. (1993) The mahogany shoot borer: prospects for control. *Forest Ecology and Management* 57, 301–328.

Newton, A.C., Leakey, R.R.B. and Mesén, J.F. (1993) Genetic variation in mahoganies: its importance, capture and utilization. *Biodiversity and Conservation* 2, 114–126.

Newton, A.C., Leakey, R.R.B., Baker, P., Ramnarine, S., Powell, W., Chalmers, K., Mathias, P.J., Alderson, P.G. and Tchoundjeu, Z. (1994) Domestication of mahoganies. In: Leakey, R.R.B. and Newton, A.C. (eds) *Tropical Trees: Potential for Domestication. Rebuilding Forest Resources.* HMSO, London, pp. 256–266.

Newton, A.C., Cornelius, J.P., Mesén, J.F. and Leakey, R.R.B. (1995) Genetic variation in apical dominance of *Cedrela odorata* seedlings in response to decapitation. *Silvae Genetica* 44, 146–150.

Wilson, J., Munro, R.C., Ingleby, K., Mason, P.A., Jefwa, J., Muthoka, P.N., Dick, J.McP. and Leakey, R.R.B. (1991) Agroforestry in semi-arid lands of Kenya – role of mycorrhizal inoculation and water retaining polymer. *Forest Ecology and Management* 45, 153–163.

Index

Note: page numbers in *italics* refer to figures, tables and boxes; those with suffix 'n' refer to footnotes.

Aboriginal communities (Australia) 119–22
Acacia trees 35
Ackworth, James 64n
Actinidia chinensis (kiwi fruit) 177
Adansonia digitata (baobab) 37, 38
Adel, Saskia den 139n
African plum/pear *see* safou
agricultural industry
    agribusiness companies 152
    dairy industry in Kenya 32
    see also commercialization; public–private partnerships (PPPs)
agricultural landscapes 161
agricultural production
    agrichemicals 15, 52, 152
    bees/beekeeping 160
    cash crops 26
        development from indigenous crops 152
        smallholders 38
        trees 148
    cut-and-carry livestock systems 32
    distribution of benefits 15

energy costs of agriculture 18
farm size 26
food production
    approaches 143–4
food security 151, 172
    agroforestry 48–9
    staple food crops 160
global 15–16
improved fallow 27, *28*, 30
    closing the yield gap 161
intensive agriculture 15, 53
    destructiveness 52
    productivity 142
    sustainability 143
irrigation 15
mechanization 15
minimum tillage 53, 143
monocultures 52–3, 177
Norfolk System 148
pesticides 15
purchasing of agricultural inputs 151
starchy food staples 147
subsistence farming 15
see also Green Revolution

Index

agriculture
    agricultural landscapes 161
    changes to feed world
        population 14–15
    climate change 145
    diversification 61, 174
    ecosystem impact 145
    energy costs 18
    environmental degradation
        21, 145
    good practices 145
    greenhouse gas production 18
    impacts 170–1
    integrated approaches 143
    intensification 138, 157
    land for 145–6
    natural capital 19
    productivity loss 19
    starchy food staples 147
    sustainability 22–3
    sustainable development 142
    water scarcity 145
    workforce 18
    see also farming systems; multi-
        functional agriculture
agroecology 51–2
    biodiversity in oil palm
        plantations 44
    carbon sequestration 44, 149, 168
    drought 13, 16, 26, 56, 61, 149
    ecological succession 52–3
    ecological sustainability 52
    experiments 60
    food chain 55
    fungal filaments 54
    hydrological cycle 56, 142, 149
    mycorrhizal fungi 54
    nutrient recycling 20–1, 30, 54
    plant disease vulnerability 130
    push–pull technology 154n
    replacement series 60
    water cycle 56, 142, 149
agroecosystems
    cocoa farming 58
    ecological succession 52
    function breakdown 146–7
    function improvement 168
    management 53
    scale 55
    stability 53
    trees 148
agroforestry 11, 24–5, 148–9
    agrichemical use 152
    alley farming 24, 26
    Asia 44, *45*
    biodiversity 44
    carbon sequestration 44, 149, 168
    cocoa farming 48
    commercial tree fallows 148–9
    community impact 173–4
    definition 52–3
    ecological benefits 57
    ecological/economic
        sustainability 52
    economic benefits 44, 46
    evergreen agriculture 29
    fertilizer trees 29, 160
    food security 49
    hedgerow intercropping 24, 26
    hedges
        baobab 38
        contour 31
        fodder banks 37
        terrace stabilization 31
        thorny for crop protection 37
        water capture 30
    hydrology 149
    Kenya 32–4
    Latin America 44, *45*
    Mediterranean region 175
    missing ingredients 175–6
    multifunctional agriculture 153,
        154, 159–60
    new crops 172–3
    Peru 46
    polycultural intensification 150
    practice 25
    profitable production 53
    relay cropping 28–9
    research teams 25
    rural poverty 17
    Rural Resource Centres
        (RRCs) 157–8, 160,
        167, 169
    scaling up 62
    science of 62, 64n
    shade crops 61
    social benefits 44

teaching in tertiary colleges/
    universities 62, 63
technology expansion 168
temperate regions 175
terminology 49n
tree cultivation in
    communities 73
water cycle 149
water harvesting role of
    trees 30, 31
Agroforestry Systems (journal) 62
agroforestry tree products (AFTPs)
    see tree products
Akyeampong, Ekow 49n
Ala, Philimon 123n
Alegre, Julio 49n
Aleurites moluccana (candlenut) 40
Allanblackia (oil-seed) 130, 136–7
Alnus acuminata 31
Ananas erectifolius (curaná) 135
andiroba (*Carapa guianensis*) 135
Anegbeh, Paul 85, 93n
Annandale, Mark 123n
applied agroecology 51–2
Asia agroforestry 40–4, 45, 46
    cinnamon 43
    damar 40, 43
    dipterocarp 40
    rubber 39, 43
    see also Sumatra agroforestry
Atangana, Alain 84, 137
Australia
    bush tucker industry 120–3
    Far North Queensland 119–20
    James Cook University
        (Cairns) 112
Avila, Marcelino 49n
Ayuk, Elias 49n
Azanza garckeana 29

Bactris gasipaes (peach palm) 47, 75
Bandy, Dale 49n
Bangor University (North Wales) 6, 8
baobab (*Adansonia digitata*) 37, 38
Barringtonia procera (cutnut) 113,
    115–16
biodiscovery 78–9
biodiversity 19
    agroforestry 44
    enhancement 168
    increase 52–3
    loss 19, 146
    Miombo woodlands 29
    oil palm plantations 44
    planned 52, 53–4, 59, 174
    retention 53
    unplanned 52, 53, 54, 59
biopiracy 127
bioprospecting 78–9
biotechnology 144
bird cherry (*Prunus avium*) 107
bitter kola (*Garcinia kola*) 69, 74, 77
bitter leaf (*Vernonia*) 162
Boehmeria nivea (ramie) 135
Boland, Doug 81n
Bolivia 48
Bongkoungou, Edouard 49n
Borlaug, Norman 16
Botelle, Andy 139n
Botha, Jenny 139n
branching and apical dominance 91
Brazil, agroforesty 46, 47–8, 49n, 181
    cocoa growing 57–60
    public–private partnership 134–5
Brennan, Eric 95
Bunt, Colin 123n
Buresh, Roland 49n
Burundi 32–3
Burundi, tree growing 33
bush mango (*Irvingia gabonensis*)
    2–3, 40
    data collection 84–5
    domestication 73, 77, 91
    fruit size variation 86, 88, 90–1
    harvesting 3
    kernels 88–9
    nutshell thickness variation 86
    thickening agent 88–9
    trait values 87
bush meat 164
bush tucker industry 120–3
    consumers 122
    Native Foods industry 120–3
    Peak Body 122–3
    producers 122
    training for indigenous
        students 121–2

Bush Tucker Summit 122
Busseola fusca (stem borer) 148

Calliandra calothyrsus (nitrogen-fixing leguminous tree) 28, 31, 32
Cameroon 83, 84, 86, 87, 89, 90, 126
    cocoa growing 39
    Food for Progress project 157–160, 162, *165–6*, 168
    medicinal products of trees 34
    plus-tree selection 127
Canarium indicum (galip nut) *see* galip nut (*Canarium indicum*)
candlenut (*Aleurites moluccana*) 40
canopy trees
    cocoa farming 57, 58, 59
    Nelder fan 58–9
    replacement series 60
Cape York (Queensland) 119–20
Carapa guianensis (andiroba) 135
carbon cycle 54
carbon dioxide emissions 54, 56, 174
carbon sequestration 44, 149, 168
Castillo, Carlos 49n
Centre for Ecology and Hydrology 11n
Chapman, Vicky 109n
Chilo partellus (cereal stem-borer) 26, 148
Chrysophyllum albidum (star apple) 74, 77
Cinderella trees 11
climate change 56
    agriculture impact 142, 145
    carbon sequestration 44, 149, 168
    minimizing 172, 174
    perennial vegetation loss 146
clones 96
cocoa farming 39, 57–60
    agroecosystems 58
    agroforestry 48, 57–8
    canopy trees 57, 58, 59
    Nelder fan 58–9
    pests 117
    planned biodiversity 59
    replacement series 60
    shade trees 57, 58, 59
    tree planting 57–8
    unplanned biodiversity 59
    witch's broom disease 57
cocoa pod borer (*Conopomorpha cramerella*) 117
Cola nitida 161
Combrinck, Adrian 139n
commercialization 125, 127
    agroforestry tree products 130–1, 134, 151, 153
    alcoholic beverage brewing 126
    Amarula liqueur 131
    business skills 159
    car manufacturing 135–6
    cassava-processing mills 162
    certification schemes 138
    Distell Corporation 131
    diversification 174
        risk-aversion 61, 174
    diversity scale 61
    herb packaging 162–3
    impacts 131, *132–3*, 133
    jam making 126
    marula beer 128
    new tree crops 173
    processing equipment 159, 162–3
    simultaneous with domestication 134
    tree products 130–1
    trading 128
    value-adding to tree products 126
    value chain 152
communities
    agroforestry impact 173–4
    empowerment 73, 164, 174
    infrastructure development 159, 163–4
    project management 159
    Rural Resource Centres 158
Conopomorpha cramerella (cocoa pod borer) 117
conservation agriculture 143
Convenient truths 172–6
Convention on Biological Diversity 73, 78
Conway, Gordon 21, 23n
copyright 127
Cordia africana 33, 46
Costa Rica 48
Coutts, Mike 108n

Craig, Nola 123n
Cribb, Julian 16, 23n, 145, 154n, 178
Cribbins, Jill 139n
Crinipellis perniciosa (fungus) 57
crop(s)
    hedges for protection 37
    leguminous nitrogen-fixing 26, 147–9
    new 172–3
    pest resistance 151
    rotation 148
    shade 61
    shade tolerant 64n
    small scale mixed cropping 130
    weeds 148
    year-on-year cropping 26
    see also food crops
crop breeding 72–3
    genetic modification 151
    investment 152
    pest resistance 151
    plant breeding for Green Revolution 15
crop yields 147, 150–1
    decline 26
    peak biological yield 147
    see also yield gap
Croton megalocarpus 33
Cryptomeria japonica 9
cultivars 95
    development 92
    manipulation 97–8
    records of development 127
Cunningham, Tony 139n
Cupressus lusitanica 33
Cupressus macrocarpa 33
curaná (*Ananas erectifolius*) 135
cutnut (*Barringtonia procera*) 113
    tree variability 115–16

Dacryodes edulis (safou) *see* safou (*Dacryodes edulis*)
dairy industry, Kenya 32
damar (*Shorea javanica*) 40, 43
deforestation 10, 15, 18, 145–6
    hydrological cycle 56
    Miombo woodlands 29
    reversal with population growth 33, 44, 46
Degrande, Ann 81n
Denovan, Debbie 109n
Department for International Development (DFID) 11
Desmodium (legume crop) 148
Diamond, Jared 22, 23n, 176, 179n
Dick, Jan 108n, 109n
Dickinson, Geoff 123n
Djimdé, Mamadou 49n
domestication 172–3
    agroforestry tree products 130–1, 134
    clones 96
    costs:benefits 177
    dogs 72
    food crops 176
    genetic diversity loss 78
    genetic selection for improvement of characteristics 76
    genetic variation 71–2, 172
    income generation 173
    mature trees 96–7, 107
    plants 65–6, 70
        definition 70
        identification of trees for 71
        socio-economic activities 66
        traditional foods 69
    plus-tree selection 72, 83–93
        community programmes 127
        cultivars 95
        genetic variability 116
    processes 125
    Rural Resource Centres (RRCs) 157–8, 160, 167, 169
    shade crops 61
    simultaneous with commercialization 134
    species numbers 177
    tree choice of farmers 73–6
    tree cultivation 73
    tree identification 71
    yield gap closure 159–60
    see also trees, domestication
Doubly Green Revolution 21, 23n
Duguma, Bahiru 81n

duku (*Lancium domesticum*) 40
durian (*Durio zibethinus*) 40

East, Ken 109n
East African Highlands 33
    medicinal products from trees 34
East New Britain (Papua New
    Guinea) 116–17
ecoagriculture 112, 143
ecological sustainability 52
ecology *see* agroecology
economic accounting 32
economics
    problems 16, 17
    supply and demand laws 130
ecosystems
    agriculture impact 145
    health 19
    tropical 54, 55
    see also agroecosystems
Emmanuel, Philippa 139n
environmental degradation 145
    agricultural impacts 21
    environmental accounting 31–2
    flooding 31, 142, 146
    fossil fuel use in agriculture 18
    groundwater pollution 20
    landslides 31, 146
    pollution 19, 20
    woodland clearing 26
environmental problems, civilization
    crashes 22
environmental services, payment
    for 149
Environmental Standard (ISO
    14062) 136
erosion 19, 142
    control measures 30–1
    deforestation 146
    nitrogen-fixing leguminous
        trees 30
eru (*Gnetum africanum*) 3, 4, 5
    harvesting 5
    packaging 162–3
    propagation 5
    shade crop 61–2
Erythrina peoppigiana (nitrogen-fixing
    leguminous tree) 46

Esslemont, Dick 11n
Eucalyptus 9
Eucalyptus grandis 33
Eucalyptus saligna 33
exploitation 127

Faidherbia albida (nitrogen-fixing
    leguminous tree) 29
fair trade 138
Fairley, Sue 123n
Fairtrade Foundation 153
Far North Queensland
    (Australia) 119–20
farmers
    building trust/partnerships 84
    diversification 61, 148–9,
        150, 164, 178
    drought concerns 56
    empowerment 73, 158, 172
    food insecurity 143, 152
    improved fallow methods 27
    improvement of lot 5
    income generation pathway 127
    indigenous timber tree
        planting 33–4
    innovative practices 25
    intellectual property rights 177
    livelihood benefits 8
    loans for agricultural inputs 163
    malnutrition 15
    marginalization 138
    motivation 19
    new technology
        dissemination 7
    partnership building with
        scientists 84–6
    poor 5, 130, 174
    poverty 19, 142, 147
        reduction 134, 152
    proportion of population 14, 26
    rights over traditional know-
        ledge 127, 133, 134, 177
    sedentary 15
    seed production/
        distribution 27–8
    smallholders 8, 11, 15, 32, 147
    subsistence 15, 26
    tree selection 90–1

tree species for domestication 73, 75, 76, *77*, 78–9, 88, 98
working with 85–6
yield gap 147
see also Rural Resource Centres (RRCs)
farming systems 51–62, *63*
   cut-and-carry livestock systems 32
   high-energy 145
   livestock 150
   low-input 145
   minimum tillage 53, 143
   monocultures 52–3, 177
   nomadic pastoralists 35, 37
   Norfolk System 148
   oil palm plantations 38, 44
   organic agriculture 143
   paddy rice fields 41, 44
   permaculture 18, 143
   scale of tree-based 175
   sedentary farming 15
      protection of crops 37
      Sahel 35–6
      Sumatra agroforestry 43
   shifting cultivation 14–15, 46
      Brazil 47
      Sumatra 42, 43
   small-scale mixed cropping 130
   subsistence farming 15, 26
   see also smallholder farming
Ferguson, Nina 93n, 109n
fertilizers, artificial 15, 19
   Green Revolution 15
   purchase 149, 150–1
flooding 31, 142, 146
fodder
   availability in Sahel 36–7
   tree products 32
fodder banks 36–7
Fondoun, Jean-Marie 81n, 93n
food(s), traditional 67–9, 173
food crisis 16, 138–9
   environmental 178–9
   global 16
food crops 26
   domestication 176
   staple 68, 160

Food for Progress project (Cameroon) 157–64, *165–6*, 166–9
   adoption rates 166–7
   business skills 159
   cassava-processing mills 162
   community project management 159
   concept 159–60
   empowerment 164
   expansion 158–9
   impacts 164, *165–6*, 166
   implementation 157–8
   income generation 161–2
   income increase 167
   infrastructure development 159, 163–4
   loans 159, 163
   marketing skills 159
   microfinance 159, 163
   new components 159
   processing equipment 159, 162–3
   Rural Resource Centres 157–8, 160, *167*, 169
   satellite nurseries 158, 160–1
   women's initiatives 161–2
   young people remaining in villages 166
food security 27, 48–9, 53, 125
   soil fertility amelioration 149, 151, 160, 171
   yield gap closure 148, 151
forest clearing, land degradation 26
Foresta, Hubert de 49n
Forse, Bill 12n
Fourmile, Seith 122
Franzel, Steve 49n, 81n
Freebody, Kylie 123n
fruit(s)
   indigenous 1–3, *4*, 5, 40, 73, 78
   nutritional value 90
   nuts/kernels 71
      Allanblackia 136–7
      baobab 38
      bush mango 3, *4*, 73, *77*, 84, *86*, 87, 88–9
      cocoa 84
      cutnut *113*, 115–16
      galip nut *114*, 116–17

fruit(s) (*continued*)
    ideotypes 88, 89
    markets 88
    marula 131, 133, 134
    medicinal products 116, 117
    njangsang 74, 77
    okari nut *114*
    safou 40, 84, 91
  shelf life 126
  taste quantification 89–90
  tree-to-tree variation in traits 79, 86–7
fruiting season 90

galip nut (*Canarium indicum*) 112, *114*
  domestication 116–17
  medicinal product 116
  medicinal properties of oil 89
  tree-to-tree variation 117
Garcinia kola (bitter kola) 69, *69*, 74, 77
Garrity, Dennis 1, 24, 49n
genetic resources
  adoption by farmers 66
  conserving 78
  genetic diversity loss with domestication 78
  sustainable use/improvement 73
Gliessman, Stephen 51
Gliricidia sepium (nitrogen-fixing leguminous tree) 28, 37, 147
global food crisis 16
globalization 13, 14
  balance with localization 138
  consumptive philosophy 21
  enlightened 138, 174
  Green Revolution benefits for developing countries 16
  impact 14
  industrialized countries 14, 16
  multinational companies 138, 174
  trade 138
Gnetum africanum (eru) *see* eru (*Gnetum africanum*)
Gore, Al 18, 23n, 170, 179n
Grace, John 93n
Green Revolution 15, 65, 142
  agricultural production 15, 142
  benefits in developing countries 16
  crop breeding investment 152
  India 6–8
  investment returns 172
  pesticides 15
Green Revolution in Africa (Gates Foundation) 21
greenhouse gases (GHGs) 18, 53
  emission reduction 174
  land degradation 56
Grevillia robusta (timber tree) 31

Haggar, Jeremy 49n
Hamzah, Aminah 109n
Hartmann, Hudson 95, 109n
Harvey, Frank 109n
hedgerow intercropping 24, 26
Helling, Marianne 123n
Herren, Hans 154n
HIV/AIDS susceptibility 69
Hoad, Steve 108n
hunger 16–18
  agricultural productivity loss 19
  alleviation 151
  prevalence 142
hydrological cycle 56, 142, 149

ideotypes 88, 89
India
  Green Revolution 6–8
indigo (*Indigofera arrecta*) 135
Indonesia, agroforests 40–4, *45*, 46
Inocarpus fagifer (Tahitian chestnut) *see* Tahitian chestnut (*Inocarpus fagifer*)
Institute of Terrestrial Ecology 11n
Institute of Tree Biology (Edinburgh) 1–2, 8–9, 11n
integrated approaches 143, 168, 171, 178
  integrated natural resources management 154
  multinational companies 174
  see also multifunctional agriculture
intellectual property rights 127, 177

Trade-Related Aspects of
   Intellectual Property
   Rights (TRIPS) 78
 see also traditional knowledge
International Assessment of
   International Knowledge,
   Science and Technology for
   Development (IAASTD) 142
   impact statements 143
   outcomes 145
   reports 144–5
International Centre for Research
   in Agroforestry (ICRAF,
   Nairobi) 11, 12n, 24–5, 30
   technology dissemination 27–8
international development
   developing countries 14, 143, 167
   industrilized countries 143
   law 127
international public goods and
   services 53
Irvingia gabonensis (bush mango)
   see bush mango (*Irvingia
   gabonensis*)

Jackson, Nick 49n
Jaenicke, Hannah 81n
Jama, Bashir 49n
James Cook University (Cairns,
   Australia) 112
Janssen, Willem 81n
Jelly, Roy 11n
Jenik, Jan 1
Johnston, Mark 123n

Kalinganire, Antoine 49n
Kalomor, Leimon 123n
Kengni, Eduard 89
Kengue, Joseph 81n, 93n
Kester, D.E. 95, 109n
Khan, Ahmed 49n
Khaya nyasica 29
Kindt, Roeland 49n
kiwi fruit (*Actinidia chinensis*) 177
Knees, Sabina 109n
Kolombangara Island (Solomon
   Islands) 113

Konuche, Paul 34
Kumba market (Cameroon) 1–3,
   4, 5, 10
Kwesiga, Freddie 49n

Laamanen, Risto 139n
Ladipo, David 81n, 93n
Laird, Sarah 139n
Lancium domesticum (duku) 40
land clearance 142
   forest/woodland 26
   see also deforestation
land degradation 18–19, 20, 21
   acid soils 28
   agriculturally induced 55–6
   crop yields 147
   forest/woodland clearing 26
   land use in tropics 146
   landslides 31, 146
   poverty association 19, 21
   rehabilitation 172
   social deprivation 151
landscape
   agricultural 161
   more people:more trees 33, 44, 46
   mosaic 55, 61
   watershed protection 53, 149, 168
Last, Fred 46
Latin America 46–8
   agroforestry 44, 45, 46–9
   Nelder fan 58–9
leguminous crops 26, 147–8
   nitrogen fixation 25, 26, 30, 157
   yield gap closure 147–9
   see also nitrogen-fixing
      leguminous trees
Lewis, Fonda 139n
livelihoods
   access to better 175
   benefits 8
   declining 20, 146
   enhanced 49, 116–17, 135, 172
      multifunctional
         agriculture 142–5,
         150, 154, 156
      yield gap closing 159–60
   meaningful benefits 137
   needs 22–3

livelihoods (*continued*)
    options 7
        diversification 164
        positive impacts 164
        Sumatran agroforestry 42–3
        unreliable strategies 26
livestock 150
    cut-and-carry systems 32
loans 163
localization, globalization balance 138
Lombard, Cyril 139n
Longman, Alan 93n, 108n, 109n, 168, 169n

Maathai, Wangari 14, 23n, 179
macadamia nut (*Macadamia ternifolia*) 177
Maghembe, Jumanne 49n
maize growing 26
    genetic modification 151
    Grevillia robusta interplanting 31
    improved fallow 27
    soil degradation 147–8
    yield gap closure 151
    yields 147–8
malnutrition 16–18
    agricultural productivity loss 19
    alleviation 151
    indigenous tree domestication 90
    prevalence 142
    smallholders in tropics 15
Mander, Myles 139n
Mangifera indica 33
Manurung, Rita 93n
Marchant, Fred 123n
marketing of tree products 125–39
    commercialization impacts 131, *132–3*, 133
    local 177
    market forces in tree domestication 76
    outlets 163
    rights over traditional knowledge 127–8
    skills 159
    small-scale 130
    smallholder households 130
    see also commercialization

marula tree (*Sclerocarya birrea*) 29, 131
    Amarula liqueur 131
    commercialization 131, 133
    domestication 134
    fruits 131
    ideotype 89
    kernels 131, 133–4
    male/female trees 134
    marula beer 128, 131
    oil extraction 133
    tree-to-tree variation 133–4
Maund, Victor 123n
McBeath, Colin 109n
McHardy, Tania 139n
McIntosh, Richard 123n
McIntyre, Beverley 154n
McNeely, Jeffrey 112
Mediterranean region 175
Melnyk, Mary 65
Meru oak (*Vitex keniensis*) 33–4
Mesén, Francisco 109n
Michon, Genevieve 49n
microfinance 159, 163
micronutrients 151, 173
Millennium Development Goals (UN) 21, 23n, 167
Miombo woodlands 29, 131
Mohammed, Hassan 109n
Mollinson, Bill 18, 23n
Moxon, John 123n
Mudge, Kenneth 95
Muller, Jillian 139n
multifunctional agriculture 144, 150, 151–2, 156–69
    agroforestry 153, *154*, 159–60
    drivers of change 171
    economic benefits of innovations 127
    infrastructure development 159, 163–4
    loans 163
    microfinance 159, 163
    missing ingredients 175–6
    multinational companies 138, 174
    push–pull technology 154n
    rural poverty 17
    Rural Resource Centres (RRCs) 157–8, 160, *167*, 169
    yield gap closure 159–60

multinational companies
  Novella Partnership 136–7
Munro, Bob 46

nangai *see* galip nut (*Canarium indicum*)
napier grass (*Pennisetum purpureum*) 32, 148
natural resources
  agricultural practices impact 145
  exploitation 18, 21, 22, 126, 156
    colonialism 14
    Green Revolution 142
  integrated management *154*
  management 53, 138, 143
  overuse 16
  sustainable development 143
  sustainable use 168, 171
  unsustainable use impact *20*
Ndam, Nouhou 64n
Ndlovu, Sibongile 139n
Ndungu, Julia 81n
Nelder fan 58–9
Nelleman, Christian 178
néré (*Parkia biglobosa*) 37, 38
Netshiluvhi, Thiambi 139n
Network for Sustainable and Diversified Agriculture (NSDA) 122
Nevenimo, Tio 123n
Newton, Adrian 10, 66, 108n
ngali *see* galip nut (*Canarium indicum*)
Niang, Amadou 49n
Nigeria, cocoa growing 39–40
nitrogen-fixing crops 26, 147–9
nitrogen-fixing leguminous trees 24, 27, 147–8
  erosion control 30
  fodder production 32
  food security enhancement 151
  germination 28
  Latin America 46
  rainfall capture/use 30
  Sahel 37
  simultaneous crop growing 28–9
  soil fertility enhancement 160
njansang (*Ricinodendron heudelotii*) 3, 4
  domestication 75, 76, 161
  harvesting 5

Nkefor, Joseph 5, 62, 64n
Nketiah, Theresa 109n
nomadic pastoralists, Sahel 35, 37
nutrition, human
  diet improvement 173
  micronutrients 151, 173
  poor and disease immunity 69
  unhealthy diet 68–9
nuts, indigenous 1–3, 4, 5, 40, 73

obeche (*Triplochiton scleroxylon*) 1, 8
  genetic variation 91
Ofori, Daniel 109n, 137
Okafor, Victoria 84
okari nut (*Terminalia kaernbachii*) 112, *114*
Oldham, Steve 123n
O'Neill, Mick 49n
Ong, Chin 49n
O'Regan, Dermot 139n
Osborne, David 123n
Ottley, Ray 109n

paddy rice fields 42, 44
Page, Tony 118–19
Pakistan 7
Papua New Guinea 116–17
  historical food crop domestication 176
Parkia biglobosa (néré) 37
participatory processes 76–8, 157, 173
Pate, Kris 139n
Patterson, Rob 49n
Pauku, Richard 109n, 113, 115–16
Pausinystalia johimbe (yohimbe) 80
peach palm (*Bactris gasipaes*) 47, 75
Peden, Don 49n
Pennistum purpureum (napier grass) 32, 148
Peprah, Theresa 137
Peru 46, 48
phytochrome 105
PhytoTrade Africa 128, 138, 153
Pinstrup-Andersen, Per 156
plant breeding, Green Revolution 15

plantations
    colonial era 14
    monocultural 177
    oil palm 38, 44
    replacement with trees 33
    rubber 38
    shade 48
Plessis, Pierre du 139n
policies
    appropriate 176
    integrated development 163
    see also sustainable development
political will 175
Pollan, Michael 65
population growth 142
    food needs 14–15
    tropical deforestation reversal 33, 46
Poulsen, Uffe 139n
poverty 17–18
    agricultural productivity loss 19
    alleviation 21, 126, 151
    definition 139n
    land degradation association 19, 21
    pathway out 128, 138
    prevalence 143
    smallholders 147
    trap 142
Poverty and Environment in the Amazon (POEMA) 135–6
predators
    top 55
    vulnerability 130
Pretty, Jules 65
Prunus africana (pygeum) 34, *35*, 79–80, 161
Prunus avium (bird cherry) 107
Pterocarpus angolensis (nitrogen-fixing leguminous trees) 29
Pterocarpus erinaceus (nitrogen-fixing leguminous trees) 37
public–private partnerships (PPPs) 134–8, 153, 174
    Daimler car manufacturing 135–6
    Mercedes car production 135, 136
    Novella Partnership 136–7
pygeum (*Prunus africana*) 34, *35*, 80
    domestication 161

rainfall 56
    drought 13, 16, 26, 56, 61, 149
    regimes in Africa 25
    runoff 146
    soil water 56
ramie (*Boehmeria nivea*) 135
Rao, Meka 49n
Reading University 8
Ricinodendron heudelotii (njangsang) 3, 4, 74, *77*, 162
Roberts, Sir William 7
Robson, Ken 123n
rooting
    cuttings 99
    genetic variation 107
    stockplant management 104, 106, 107, 108, 117
    success determination 107–8
    success maximization 101
rural development
    integrated approaches 143, 168
    multifunctional agriculture 141, 143, 144, 153, *154*
    see also agroforestry; social development; sustainable development
Rural Resource Centres (RRCs) 157–8, 160, *167*, 169
    village nurseries 158

Sachs, Jeffery 138, 139n
safou (*Dacryodes edulis*) 40
    data collection 84
    domestication 73–4, 77, 161
        by local communities 90
    fruit crop 79
    fruit size variation 86, *87*, 90–1
    nutshell thickness
        variation 86
    seasonality 126
    seedless fruits 89
    trait values 87
Sahel 35–8
    fodder availability 37
    nitrogen-fixing leguminous trees 37
    parklands 36–37
    soumbala 38

Sam, Chanel 123n
Sanchez, Pedro 24, 30, 49n
sandalwood (*Santalum austrocaledonicum*) 117–19
   domestication 118
   essential oils 89, 119
   oil distillation 118
   parasitic habit 118
   sampling 118–19
   tree-to-tree variation 119
sandalwood (*Santalum lanceolatum*) 119–20
Scherr, Sara 112
Schreckenberg, Kate 79
Sclerocarya birrea (marula) *see* marula tree (*Sclerocarya birrea*)
Scoones, Ian 65
Scott-Rimington, Tracy 123n
Seale-Hayne Agricultural College (Devon) 5–6
Sesbania sesban (nitrogen-fixing leguminous tree) 27, *28*, 147
Shackleton, Charlie 134, 139n
Shackleton, Sheona 49n, 139n
shade 106
   agroecosystems 25, 28, 39
   cocoa growing 57–60, 73, 148
   eru growing 3, 5, 62
   plantation crops 49
   vegetative propagation 99, 105–6, 108, 117
Shapland, Nick 11n
shea nut (*Vitellaria paradoxa*) 36, 37
Shiembo, Patrick 1–2, 5, 109n
   eru domestication 62
Shorea javanica (damar) 40, 43
Simons, Tony 49n, 81
Sinclair, Upton 170
smallholder farming 8, 11, 15, 32, 147
   cash crops 38
   diversification 150
   low-input farming 145
   malnutrition in tropics 15
   marketing of tree products 130
   poverty 147
   tree crops 148
   see also food security; yield gap
Smithson, Paul 49n

social development
   aid 11
   civilization crashes 22
   colonialism impact on Africa 14
   conflict 16
   consumption orientation of society 22, 153
   culture loss 127–8
   developing countries
      Green Revolution benefits 16
      rural population 14
      social support 14
   European colonists 13
   food insecurity 143
   industrialized countries 14, 16
   innovation protection 177
   loans 163
   microfinance 159, 163
   poor governance 16
   retention of young people in villages 166, 173
   social consequences 10
   social deprivation 20, 21
      link with land degradation 151
   tree domestication 90–1
soil fertility
   acidification 38
   artificial fertilizers 15, 19
      purchase 149, 150–1
   carbon storage 168
   decline 26
   enhancement by fertilizer trees 29, 160
   fertile 18
   fertilizer trees 29, 160
   food security 149, 151, 160, 172
   improved fallow 30
   improvement 19–21
      technologies 172
   loss 146–7
   manure 19
   nitrogen 147–8
   nitrogen levels 147–8, 151
   nutrient cycle 54
   overexploitation 146
   restoration 148
   runoff 146
   temperate climate 55

soil fertility (*continued*)
:   tropical ecosystems 55
    West African humid forest zone 38
    see also erosion
Solomon Islands 113, *113*, 115–16
species
:   area of origin 128, 130
    numbers domesticated 177
star apple (*Chrysophyllum albidum*) 74, 77
Storeton-West, Richard 109n
Striga hermonthica (witchweed) 26, 148, 151
Strychnos cocculoides 29
Stubbs, Simon 11n
Sullivan, Caroline 139n
Sumatra agroforestry 40–4, *45*, 46
:   economic benefits 44
    food crops 43
    market opportunities 43
    productivity 41–2
    sedentary agriculture 43
    social benefits 44
sustainable development 21, 139n, 141–3
:   agriculture contribution 142
    enlightened globalization 138, 174
    food crisis 16, 138–9
        environmental 178
    hunger 16–18, 19
        alleviation 151
        prevalence 143
    rural poverty 17
    sustainable production 138
    see also agroforestry; Convenient truths; malnutrition; multifunctional agriculture
Swift, Mike 1

Tahitian chestnut (*Inocarpus fagifer*) 113, *115*
Tate, Hanington 123n
Taylor, Colin 123n
Tchoundjeu, Zac 49n, 81n, 93n, 109n, 137, 156, 157, 166
temperate regions
:   agroforestry 175
    soil fertility 55

Tephrosia vogelii (nitrogen-fixing leguminous tree) 27, 147
Terminalia kaernbachii (okari nut) 113, *114*
Theobroma cacao (cocoa) *see* cocoa farming
thunder god vine (*Tripterygium wilfordii*) 80
Trade-Related Aspects of Intellectual Property Rights (TRIPS) 78
trading, tree products 128
traditional foods 67–8, 173
:   access to 69
    cultivation through agroforestry 69
    domestication 69
    nutritional value 67–8
traditional knowledge 173
:   farmers as beneficiaries/ guardians 78
    farmers rights over 127, 133, 134, 177
    ju-ju symbols 85
    loss 127
    traditional medicine 69, 79, 173
Tranent, Rob 123n
tree(s)
:   apical dominance and branching architecture 91
    breeding
        generation time 91, 96
        progeny 96
    domestication
        choice by farmers 73–6
        pathway out of poverty 128
        woody plant revolution 11
    fruiting season 90
    selection 83–93
        ideotypes 88–90
        shelf life of fruits 126
        tree-to-tree variation in fruit traits 79, 86–7
    see also nitrogen-fixing leguminous trees; vegetative propagation
tree nurseries
:   commercial 160–1
    community 31, 98
    income generation 161, 164, 166

plant multiplication 160–1
satellite 158, 160–1
tree planting on farms 160–1
tree products 10, 32, 80–1
    aphrodisiacs 80, 131
    benefits maximization 134
    car manufacture 135–6
    commercial 128, *129*
    commercialization 130–1, 134, 151, 153
    common property extractive resources 81
    contraceptives 80
    domestication 130–1, 134
    famine foods 10
    fodder 32
    fuelwood 34–5
    high-value indigenous timbers 33–4
    honey 160
    income source 149
    marketable 53–4, 61
    medicinal products 34, *35*, 69, 79–80, 89
    processed 126
    quality variability 66
    seasonality 126
    soumbala 38
    trading 128
    uniformity lack 66
    value adding 126
    variability 66, *125*
    see also marketing of tree products
Trees of Life initiative 69, 128, 130, 138, 178
Tribe, Derek 83, 125
Triplochiton scleroxylon (obeche) 1, 8, 10, 91
Tripterygium wilfordii (thunder god vine) 80
tropical forest ecosystems 54–5
tropical rainforest, nutrient recycling 54
Tropical Trees: the Potential for Domestication and the Rebuilding of Forest Resources (conference) 10–11, 66
Tudge, Colin 141
Tungon, Joseph 123n

Unilever plc 136–7
Usoro, Cecilia 84

van Damme, P. 49n
van Noordwijk, Meine 49n
Vangueria infausta 29
Vanuatu 117–19
vegetative propagation 72, *92*, 93, 95–108, 172
    auxins 101
    budding 96
    clones 95
    cuttings 97, 103
    factors determining success 98–101, *102*, 103–7
    grafting 96–7
        techniques 97
    irradiance 106
    light 105–6
    marcotting *92*, 96–9, *98*
    mature trees and tissues 97, 107
        precocious flowering/short stature 97–8
    non-mist propagators 100–1, *102*, 117
    nutrients 106
    phytochrome 105
    post-severance treatments 101, 103
    propagation environment 100–1, *102*
    red:far-red light ratio 105–6
    root growth 100
    rooting
        cuttings 99
        genetic variation 107
        success determination 107–8
        success maximization 101
    rooting hormones 101
    rooting medium 101, *102*
    shade 97, 105–6, 108, 117
    stockplant environment 105–7
    stockplant factors 104–5
    stomata closure 99, 103
    techniques 96–8
    tree growing in Burundi 32–3
    water stress 99, 103
Vernonia (bitter leaf) 162

Viera, Tom 123n
Vietnam
    multifunctional agricultural landscape *149*
    replacement series 60
Vitellaria paradoxa (shea nut) 36, 37
Vitex keniensis (Meru oak) 33–4

Wakhungu, Judi 154n
Waruhui, Kijo 84
water resources
    for agriculture 19
        scarcity impact 145
    misuse 18
    overexploitation 146
Watson, Bob 154n
Waycott, Michelle 120
Weber, John 50n
Weed Research Organization (Kidlington, Oxford) 8
weeds 148, 151
    vulnerability 130
Wiersum, Freerk 112
witch's broom disease of cocoa 57
witchweed (*Striga hermonthica*) 26, 148, 151
women's initiatives 161–2
    income generation 173
    loans 163

woodland clearing, land degradation 26
woodland savannah 29
World Agroforestry Centre 12n, 157
World Summit on Sustainable Development 21, 73, 145
World Trade Organization (WTO) Agreement on Trade-Related Aspects of Intellectual Property Rights (TRIPS) 78
Wynberg, Rachel 139n

yield gap 146, 147
    closure 150, 151
        with leguminous crops 147–8, 149
        multifunctional agriculture 159–60
        with tree domestication and commercialization 147–8, 149, 151, 159–60, 173
yohimbe (*Pausinystalia johimbe*) 80
Young, Tony 49n

Zambia 27, 29
Zimbabwe 29
Zizyhus mauritania (hedging) 37